Problem Books in Mathematics

Series editor:

Peter Winkler
Department of Mathematics
Dartmouth College
Hanover, NH 03755
USA

More information about this series at http://www.springer.com/series/714

Edward J. Barbeau

University of Toronto Mathematics Competition (2001–2015)

 Springer

Edward J. Barbeau
University of Toronto
Toronto, Ontario, Canada

ISSN 0941-3502   ISSN 2197-8506   (electronic)
Problem Books in Mathematics
ISBN 978-3-319-28104-9   ISBN 978-3-319-28106-3   (eBook)
DOI 10.1007/978-3-319-28106-3

Library of Congress Control Number: 2016930169

Printed on acid-free paper

This Springer imprint is published by Springer Nature
The registered company is Springer International Publishing AG Switzerland

Preface

Students from the University of Toronto have participated in the annual Putnam Competition since its inception in 1938. Many of them have competed with distinction, but others have not fared so well. Because of the way in which the Putnam scripts are marked, it is necessary to make significant progress on a problem to earn one or two points.

In 2001, I started a mathematics competition for undergraduates at the University of Toronto to offer the experience of solving problems under a time constraint. My goal was to provide a paper that would be more widely accessible while maintaining some challenge and interest. A secondary goal was to produce a set of problems that might find a place in regular college and university courses. So, I have collected them here in the hope that teachers and students may find them useful.

The problems have many sources. Some are original; some are side results in research papers; some have been contributed; some have appeared in the *American Mathematical Monthly*, and some, of unknown provenance, have been lying in folders in my office. Each problem is graded out of 10, with opportunity for part marks.

So that students have some choice, there are eight to ten problems on each paper. I expected that students normally should attempt no more than five, and to enforce this, they were told that only their best five problems would be marked unless the score on these five was at least 30. Inevitably, there were students who correctly solved more than five.

The first two years of college mathematics should provide sufficient background. Students should have knowledge of secondary level algebra and geometry, elementary linear algebra, basic number theory, elementary combinatorics, techniques of single variable calculus, as well as fundamental results on limits, continuous functions, and convergence of sequences and series.

In Chap. 1, I present the questions as they were given. However, the solutions are in the following chapters, arranged according to their subject matter in chronological order. Where a problem can be classified in more than one way, I have provided the statement of the problem in each relevant chapter, but the solution appears in only one chapter, indicated by a number in parentheses at the beginning of its statement in Chap. 1. At the beginning of some solutions, there is a name in square brackets of either a competitor or a colleague, to acknowledge the source of an approach to the problem. However, I take full responsibility for the actual presentation.

The book concludes with two appendices. The first establishes notation and conventions and lists definitions and results that might not be readily available to every student. The second lists the names of students who performed well in the competition.

I am grateful to my colleague, Victor Ivrii, for advice on diagrams; to graduate student Tyler Holden for valuable advice in solving technical anomalies; to Emile LeBlanc and Marco De la Cruz-Heredia, who look after the computer systems in the mathematics department, for technical help; to graduate students Ian Greig and Samer Seraj for checking over some of the solutions; and to my wife, Eileen, for her sharp eyes and continuous support. The problems from the American Mathematical Monthly appear with the permission, freely given, of the Mathematical Association of America.

Toronto, Canada   Edward J. Barbeau

Contents

CHAPTER 1

Problems of the Contests

Except for the first contest, for which 3 h was allotted, the time allowed for each contest was $3\frac{1}{2}$ h.

The chapter in which the solution appears is given in parentheses at the beginning of the problem.

Sunday, March 18, 2001

1. (*3*) Let $a, b, c > 0$, $a < bc$ and $1 + a^3 = b^3 + c^3$. Prove that $1 + a < b + c$.

2. (*5*) Let $O = (0,0)$ and $Q = (1,0)$. Find the point P on the line with equation $y = x + 1$ for which the angle OPQ is a maximum.

3. (*10*) (a) Consider the infinite integer lattice in the plane (i.e., the set of points with integer coordinates) as a graph, with the edges being the lines of unit length connecting nearby points. What is the minimum number of colours that can be used to colour all the vertices and edges of this graph, so that
   (i) each pair of adjacent vertices gets two distinct colours;
   (ii) each pair of edges that meet at a vertex gets two distinct colours; and
   (iii) an edge is coloured differently than either of the two vertices at the ends?
   (b) Extend this result to lattices in real n-dimensional space.

4. (*7*) Let $\mathbf{V}$ be the vector space of all continuous real-valued functions defined on the open interval $(-\pi/2, \pi/2)$, with the sum of two functions and the product of a function and a real scalar defined in the usual way.
   (a) Prove that the set $\{\sin x, \cos x, \tan x, \sec x\}$ is linearly independent.
   (b) Let $\mathbf{W}$ be the linear space generated by the four trigonometric functions given in (a), and let T be the linear transformation

© Springer International Publishing Switzerland 2016
E.J. Barbeau, *University of Toronto Mathematics Competition (2001–2015)*, Problem Books in Mathematics,
DOI 10.1007/978-3-319-28106-3_1

determined on **W** into **V** by $T(\sin x) = \sin^2 x$, $T(\cos x) = \cos^2 x$, $T(\tan x) = \tan^2 x$ and $T(\sec x) = \sec^2 x$. Determine a basis for the kernel of T.

5. (2) Let n be a positive integer and x a real number not equal to a nonnegative integer. Prove that

$$\frac{n}{x} + \frac{n(n-1)}{x(x-1)} + \frac{n(n-1)(n-2)}{x(x-1)(x-2)} + \cdots + \frac{n(n-1)(n-2)\cdots 1}{x(x-1)(x-2)\cdots(x-n+1)}$$
$$= \frac{n}{x-n+1}.$$

6. (4) Prove that, for each positive integer n, the series

$$\sum_{k=1}^{\infty} \frac{k^n}{2^k}$$

converges to twice an odd integer not less than $(n+1)!$.

7. (6) Suppose that $x \geq 1$ and that $x = \lfloor x \rfloor + \{x\}$, where $\lfloor x \rfloor$ is the greatest integer not exceeding x and the fractional part $\{x\}$ satisfies $0 \leq x < 1$. Define

$$f(x) = \frac{\sqrt{\lfloor x \rfloor} + \sqrt{\{x\}}}{\sqrt{x}}.$$

(a) Determine the supremum, i.e., the least upper bound, of the values of $f(x)$ for $1 \leq x$.

(b) Let $x_0 \geq 1$ be given, and for $n \geq 1$, define $x_n = f(x_{n-1})$. Prove that $\lim_{n \to \infty} x_n$ exists.

8. (2) A regular heptagon (polygon with seven equal sides and seven equal angles) has diagonals of two different lengths. Let a be the length of a side, b be the length of a shorter diagonal and c be the length of a longer diagonal of a regular heptagon (so that $a < b < c$). Prove ONE of the following relationships:

$$\frac{a^2}{b^2} + \frac{b^2}{c^2} + \frac{c^2}{a^2} = 6$$

or

$$\frac{b^2}{a^2} + \frac{c^2}{b^2} + \frac{a^2}{c^2} = 5.$$

Saturday, March 9, 2002

1. (8) Let A, B, C be three pairwise orthogonal faces of a tetrahedron meeting at one of its vertices and having respective areas a, b, c. Let the face D opposite this vertex have area d. Prove that

$$d^2 = a^2 + b^2 + c^2.$$

2. (2) Angus likes to go to the movies. On Monday, standing in line, he noted that the fraction x of the line was in front of him, while $1/n$ of the line was behind him. On Tuesday, the same fraction x of the line was in front of him, while $1/(n+1)$ of the line was behind him. On Wednesday, the same fraction x of the line was in front of him, while $1/(n+2)$ of the line was behind him. Determine a value of n for which this is possible.

3. *(11)* In how many ways can the rational $2002/2001$ be written as the product of two rationals of the form $(n+1)/n$, where n is a positive integer?

4. *(5)* Consider the parabola of equation $y = x^2$. The normal is constructed at a variable point P and meets the parabola again in Q. Determine the location of P for which the arc length along the parabola between P and Q is minimized.

5. *(6)* Let n be a positive integer. Suppose that f is a function defined and continuous on $[0,1]$ that is differentiable on $(0,1)$ and satisfies $f(0) = 0$ and $f(1) = 1$. Prove that, there exist n [distinct] numbers x_i $(1 \le i \le n)$ in $(0,1)$ for which

$$\sum_{i=1}^{n} \frac{1}{f'(x_i)} = n.$$

6. *(3)* Let $x, y > 0$ be such that $x^3 + y^3 \le x - y$. Prove that $x^2 + y^2 \le 1$.

7. *(7)* Prove that no vector space over $\mathbb{R}$ is a finite union of proper subspaces.

8. *(7)* (a) Suppose that P is an $n \times n$ nonsingular matrix and that u and v are column vectors with n components. The matrix $v^t P^{-1} u$ is 1×1, and so can be identified with a scalar. Suppose that its value is not equal to -1. Prove that the matrix $P + uv^t$ is nonsingular and that

$$(P + uv^t)^{-1} = P^{-1} - \frac{1}{\alpha} P^{-1} uv^t P^{-1}$$

where v^t denotes the transpose of v and $\alpha = 1 + v^t P^{-1} u$.

(b) Explain the situation when $\alpha = 0$.

9. *(10)* A sequence whose entries are 0 and 1 has the property that, if each 0 is replaced by 01 and each 1 by 001, then the sequence remains unchanged. Thus, it starts out as $010010101001\ldots$. What is the 2002th term of the sequence?

Sunday, March 16, 2003

1. *(4)* Evaluate

$$\sum_{n=1}^{\infty} \arctan\left(\frac{2}{n^2}\right).$$

2. *(3)* Let a, b, c be positive real numbers for which $a + b + c = abc$. Prove that

$$\frac{1}{\sqrt{1 + a^2}} + \frac{1}{\sqrt{1 + b^2}} + \frac{1}{\sqrt{1 + c^2}} \leq \frac{3}{2}.$$

3. *(5)* Solve the differential equation

$$y'' = yy'.$$

4. *(4)* Show that the positive integer n divides the integer nearest to

$$\frac{(n+1)!}{e}.$$

5. *(6)* For $x > 0$, $y > 0$, let $g(x, y)$ denote the minimum of the three quantities, x, $y + 1/x$ and $1/y$. Determine the maximum value of $g(x, y)$ and where this maximum is assumed.

6. *(10)* A set of n lightbulbs, each with an *on-off* switch, numbered $1, 2, \ldots, n$ are arranged in a line. All are initially off. Switch 1 can be operated at any time to turn its bulb on or off. Switch 2 can turn bulb 2 on or off if and only if bulb 1 is off; otherwise, it does not function. For $k \geq 3$, switch k can turn bulb k on or off if and only if bulb $k - 1$ is off and bulbs $1, 2, \ldots, k - 2$ are all on; otherwise it does not function.

   (a) Prove that there is an algorithm that will turn all of the bulbs on.

   (b) If x_n is the length of the shortest algorithm that will turn on all n bulbs when they are initially off, determine the largest prime divisor of $3x_n + 1$ when n is odd.

7. *(6)* Suppose that the polynomial $f(x)$ of degree $n \geq 1$ has all real roots and that $\lambda > 0$. Prove that the set $\{x \in \mathbb{R} : |f(x)| \leq \lambda |f'(x)|\}$ is a finite union of closed intervals whose total length is equal to $2n\lambda$.

8. *(7)* Three $m \times n$ matrices A, B and $A + B$ have rank 1. Prove that either all the rows of A and B are multiples of one and the same vector, or that all of the columns of A and B are multiples of one and the same vector.

9. *(5)* Prove that the integral

$$\int_0^\infty \frac{\sin^2 x}{\pi^2 - x^2} \, dx$$

exists and evaluate it.

10. *(9)* Let G be a finite group of order n. Show that n is odd if and only if each element of G is a square.

Sunday, March 14, 2004

1. (2) Prove that, for any nonzero complex numbers z and w,

$$(|z| + |w|)\left|\frac{z}{|z|} + \frac{w}{|w|}\right| \leq 2|z + w|.$$

2. (4) Prove that

$$\int_0^1 x^x\, dx = 1 - \frac{1}{2^2} + \frac{1}{3^3} - \frac{1}{4^4} + \frac{1}{5^5} + \cdots.$$

3. (11) Suppose that u and v are positive integer divisors of the positive integer n and that $uv < n$. Is it necessarily so that the greatest common divisor of n/u and n/v exceeds 1?

4. (3) Let n be a positive integer exceeding 1. How many permutations $\{a_1, a_2, \ldots, a_n\}$ of $\{1, 2, \ldots, n\}$ are there which maximize the value of the sum

$$|a_2 - a_1| + |a_3 - a_2| + \cdots + |a_{i+1} - a_i| + \cdots + |a_n - a_{n-1}|$$

over all permutations? What is the value of this maximum sum?

5. (7) Let A be a $n \times n$ matrix with determinant equal to 1. Let B be the matrix obtained by adding 1 to every entry of A. Prove that the determinant of B is equal to $1 + s$, where s is the sum of the n^2 entries of A^{-1}.

6. (5) Determine

$$\left(\int_0^1 \frac{dt}{\sqrt{1 - t^4}}\right) \div \left(\int_0^1 \frac{dt}{\sqrt{1 + t^4}}\right).$$

7. (2) Let a be a parameter. Define the sequence $\{f_n(x) : n = 0, 1, 2, \ldots\}$ of polynomials by

$$f_0(x) \equiv 1$$

and

$$f_{n+1}(x) = x f_n(x) + f_n(ax)$$

for $n \geq 0$.

(a) Prove that, for all n, x,

$$f_n(x) = x^n f_n(1/x).$$

(b) Determine a formula for the coefficient of x^k ($0 \leq k \leq n$) in $f_n(x)$.

8. (7) Let V be a complex n-dimensional inner product space. Prove that

$$|u|^2|v|^2 - \frac{1}{4}|u - v|^2|u + v|^2 \leq |(u, v)|^2 \leq |u|^2|v|^2.$$

9. (*8*) Let $ABCD$ be a convex quadrilateral for which all sides and diagonals have rational length and AC and BD intersect at P. Prove that AP, BP, CP, DP all have rational length.

Saturday, March 12, 2005

1. (*5*) Show that, if $-\pi/2 < \theta < \pi/2$, then

$$\int_0^\theta \log(1 + \tan\theta \tan x)\, dx = \theta \log \sec\theta.$$

2. (*6*) Suppose that f is continuously differentiable on $[0,1]$ and that $\int_0^1 f(x)dx = 0$. Prove that

$$2\int_0^1 f(x)^2\, dx \le \int_0^1 |f'(x)|\, dx \cdot \int_0^1 |f(x)|\, dx.$$

3. (*7*) How many $n \times n$ invertible matrices A are there for which all the entries of both A and A^{-1} are either 0 or 1?

4. (*7*) Let a be a nonzero real and $\mathbf{u}$ and $\mathbf{v}$ be real 3-vectors. Solve the equation

$$2a\mathbf{x} + (\mathbf{v} \times \mathbf{x}) + \mathbf{u} = \mathbf{O}$$

for the vector $\mathbf{x}$.

5. (*6*) Let $f(x)$ be a polynomial with real coefficients, evenly many of which are nonzero, which is *palindromic*. This means that the coefficients read the same in either direction, i.e. $a_k = a_{n-k}$ if $f(x) = \sum_{k=0}^n a_k x^k$ or, alternatively, $f(x) = x^n f(1/x)$, where n is the degree of the polynomial. Prove that $f(x)$ has at least one root of absolute value 1.

6. (*9*) Let G be a subgroup of index 2 contained in S_n, the group of all permutations of n elements. Prove that $G = A_n$, the alternating group of all even permutations.

7. (*11*) Let $f(x)$ be a nonconstant polynomial that takes only integer values when x is an integer, and let P be the set of all primes that divide $f(m)$ for at least one integer m. Prove that P is an infinite set.

8. (*7*) Let $AX = B$ represent a system of m linear equations in n unknowns, where $A = (a_{ij})$ is an $m \times n$ matrix, $X = (x_1, \ldots, x_n)^{\mathbf{t}}$ is an $n \times 1$ vector and $B = (b_1, \ldots, b_m)^{\mathbf{t}}$ is an $m \times 1$ vector. Suppose that there exists at least one solution for $AX = B$. Given $1 \le j \le n$, prove that the value of the jth component is the same for every solution X of $AX = B$ if and only if the rank of A is decreased if the jth column of A is removed.

9. (*6*) Let S be the set of all real-valued functions that are defined, positive and twice continuously differentiable on a neighbourhood of 0. Suppose that a and b are real parameters with $ab \neq 0$, $b < 0$. Define operators from S to $\mathbb{R}$ as follows:

$$A(f) = f(0) + af'(0) + bf''(0);$$

$$G(f) = \exp A(\log f).$$

(a) Prove that $A(f) \leq G(f)$ for $f \in S$;

(b) Prove that $G(f + g) \leq G(f) + G(g)$ for $f, g \in S$;

(c) Suppose that H is the set of functions in S for which $G(f) \leq f(0)$. Give examples of nonconstant functions, one in H and one not in H. Prove that, if $\lambda > 0$ and $f, g \in H$, then λf, $f + g$ and fg all belong to H.

10. (10) Let n be a positive integer exceeding 1. Prove that, if a graph with $2n+1$ vertices has at least $3n+1$ edges, then the graph contains a circuit (i.e., a closed non-self-intersecting chain of edges whose terminal point is its initial point) with an even number of edges. Prove that this statement does not hold if the number of edges is only $3n$.

Sunday, March 12, 2006

1. (10) (a) Suppose that a 6×6 square grid of unit squares (chessboard) is tiled by 1×2 rectangles (dominoes). Prove that it can be decomposed into two rectangles, tiled by disjoint subsets of the dominoes.

(b) Is the same thing true for an 8×8 array?

2. (7) Let $\mathbf{u}$ be a unit vector in $\mathbb{R}^3$ and define the operator P by $P(\mathbf{x}) = \mathbf{u} \times \mathbf{x}$ for $\mathbf{x} \in \mathbb{R}^3$ (where $\times$ denotes the cross product).

(a) Describe the operator $I + P^2$.

(b) Describe the action of the operator $I + (\sin \theta)P + (1 - \cos \theta)P^2$.

3. (2) Let $p(x)$ be a polynomial of positive degree n with n distinct real roots $a_1 < a_2 < \cdots < a_n$. Let b be a real number for which $2b < a_1 + a_2$. Prove that

$$2^{n-1}|p(b)| \geq |p'(a_1)(b - a_1)|.$$

4. (8) Two parabolas have parallel axes and intersect in two points. Prove that their common chord bisects the segment whose endpoints are the points of contact of their common tangent.

5. (10) Suppose that you have a 3×3 grid of squares. A *line* is a set of three squares in the same row, the same column or the same diagonal; thus, there are eight lines.

Two players A and B play a game. They take alternate turns, A putting a 0 in any unoccupied square of the grid and B putting a 1. The first player is A, and the game cannot go on for more than nine moves. (The play is similar to noughts and crosses, or tic-tac-toe.) A move is *legitimate* if it does not result in two lines of squares being filled in with different sums. The winner is the last player to make a legitimate move.

(For example, if there are three 0s down the diagonal, then B can place a 1 in any vacant square provided it completes no other line, for then the sum would differ from the diagonal sum. If there are two zeros at the top of the main diagonal and two ones at the left of the bottom line, then the lower right square cannot be filled by either player, as it would result in two lines with different sums.)

(a) What is the maximum number of legitimate moves possible in a game?

(b) What is the minimum number of legitimate moves possible in a game that would not leave a legitimate move available for the next player?

(c) Which player has a winning strategy? Explain.

6. (6) Suppose that k is a positive integer and that

$$f(t) = a_1 e^{\lambda_1 t} + a_2 e^{\lambda_2 t} + \cdots + a_k e^{\lambda_k t}$$

where $a_1, \ldots, a_k, \lambda_1, \ldots, \lambda_k$ are real numbers with $\lambda_1 < \lambda_2 < \cdots < \lambda_k$. Prove that $f(t) = 0$ has finitely many real solutions. What is the maximum number of solutions possible, as a function of k?

7. (7) Let A be a real 3×3 invertible matrix for which the sums of the rows, columns and two diagonals are all equal. Prove that the rows, columns and diagonal sums of A^{-1} are all equal.

8. (6) Let $f(x)$ be a real function defined and twice differentiable on an open interval containing $[-1, 1]$. Suppose that $0 < \alpha \le \gamma$ and that $|f(x)| \le \alpha$ and $|f''(x)| \le \gamma$ for $-1 \le x \le 1$. Prove that

$$|f'(x)| \le 2\sqrt{\alpha\gamma}$$

for $-1 \le x \le 1$. (Part marks are possible for the weaker inequality $|f'(x)| \le \alpha + \gamma$.)

9. (2) A high school student asked to solve the surd equation

$$\sqrt{3x - 2} - \sqrt{2x - 3} = 1$$

gave the following answer: *Squaring both sides leads to*

$$3x - 2 - 2x - 3 = 1$$

so $x = 6$. The answer is, in fact, correct.

Show that there are infinitely many real quadruples (a, b, c, d) for which this method leads to a correct solution of the surd equation

$$\sqrt{ax - b} - \sqrt{cx - d} = 1.$$

10. (*8*) Let P be a planar polygon that is not convex. The vertices can be classified as either *convex* or *concave* according as to whether the angle at the vertex is less than or greater than $180°$ respectively. There must be at least two convex vertices. Select two consecutive convex vertices (i.e., two interior angles less than $180°$ for which all interior angles in between exceed $180°$) and join them by a segment. Reflect the edges between these two convex angles in the segment to form along with the other edges of P a polygon P_1. If P_1 is not convex, repeat the process, reflecting some of the edges of P_1 in a segment joining two consecutive convex vertices, to form a polygon P_2. Repeat the process. Prove that, after a finite number of steps, we arrive at a polygon P_n that is convex.

Sunday, March 11, 2007

1. (*10*) An $m \times n$ rectangular array of distinct real numbers has the property that the numbers in each row increase from left to right. The entries in each column, individually, are rearranged so that the numbers in each column increase from top to bottom. Prove that in the final array, the numbers in each row will increase from left to right.

2. (*11*) Determine distinct positive integers a, b, c, d, e such that the five numbers a, b^2, c^3, d^4, e^5 constitute an arithmetic progression. (The difference between adjacent pairs is the same.)

3. (*10*) Prove that the set $\{1, 2, \ldots, n\}$ can be partitioned into k subsets with the same sum if and only if k divides $\frac{1}{2}n(n+1)$ and $n \geq 2k - 1$.

4. (*6*) Suppose that $f(x)$ is a continuous real-valued function defined on the interval $[0, 1]$ that is twice differentiable on $(0, 1)$ and satisfies (i) $f(0) = 0$ and (ii) $f''(x) > 0$ for $0 < x < 1$.
(a) Prove that there exists a number a for which $0 < a < 1$ and $f'(a) < f(1)$;
(b) Prove that there exists a unique number b for which $a < b < 1$ and $f'(a) = f(b)/b$.

5. (*6*) For $x \leq 1$ and $x \neq 0$, let

$$f(x) = \frac{-8[1 - (1 - x)^{1/2}]^3}{x^2}.$$

(a) Prove that $\lim_{x \to 0} f(x)$ exists. Take this as the value of $f(0)$.
(b) Determine the smallest closed interval that contains all values assumed by $f(x)$ on its domain.
(c) Prove that $f(f(f(x))) = f(x)$ for all $x \leq 1$.

6. (*4*) Let $h(n)$ denote the number of finite sequences $\{a_1, a_2, \ldots, a_k\}$ of positive integers exceeding 1 for which $k \geq 1$, $a_1 \geq a_2 \geq \cdots \geq a_k$

and $n = a_1 a_2 \cdots a_k$. (For example, if $n = 20$, there are four such sequences $\{20\}$, $\{10, 2\}$, $\{5, 4\}$ and $\{5, 2, 2\}$, and $h(20) = 4$.)

Prove that

$$\sum_{n=1}^{\infty} \frac{h(n)}{n^2} = 1.$$

7. (*7*) Find the Jordan canonical form of the matrix $\mathbf{u}\mathbf{v}^t$ where $\mathbf{u}$ and $\mathbf{v}$ are column vectors in $\mathbb{C}^n$.

8. (*8*) Suppose that n points are given in the plane, not all collinear. Prove that there are at least n distinct straight lines that can be drawn through pairs of the points.

9. (*11*) Which integers can be written in the form

$$\frac{(x + y + z)^2}{xyz}$$

where x, y, z are positive integers?

10. (*5*) Solve the following differential equation

$$2y' = 3|y|^{1/3}$$

subject to the initial conditions

$$y(-2) = -1 \qquad \text{and} \qquad y(3) = 1.$$

Your solution should be everywhere differentiable.

Sunday, March 9, 2008

1. (*8*) Three angles of a heptagon (7-sided polygon) inscribed in a circle are equal to $120°$. Prove that at least two of its sides are equal.

2. (*6*) (a) Determine a real-valued function g defined on the real numbers that is decreasing and for which $g(g(x)) = 2x + 2$.

(b) Prove that there is no real-valued function f defined on the real numbers that is decreasing and for which $f(f(x)) = x + 1$.

3. (*4*) Suppose that a is a real number and the sequence $\{a_n\}$ is defined recursively by $a_0 = a$ and

$$a_{n+1} = a_n(a_n - 1)$$

for $n \geq 0$. Find the values of a for which the sequence $\{a_n\}$ converges.

4. (*2*) Suppose that u, v, w, z are complex numbers for which $u + v + w + z = u^2 + v^2 + w^2 + z^2 = 0$. Prove that

$$(u^4 + v^4 + w^4 + z^4)^2 = 4(u^8 + v^8 + w^8 + z^8).$$

5. (*7*) Suppose that $a, b, c \in \mathbf{C}$ with $ab = 1$. Evaluate the determinant of

$$
\begin{pmatrix}
c & a & a^2 & \cdots & a^{n-1} \\
b & c & a & \cdots & a^{n-2} \\
b^2 & b & c & \cdots & a^{n-3} \\
\vdots & \vdots & \vdots & & \vdots \\
b^{n-1} & b^{n-2} & & \cdots & c
\end{pmatrix}.
$$

6. (*10*) 2008 circular coins, possibly of different diameters, are placed on the surface of a flat table in such a way that no coin is on top of another coin. What is the largest number of points at which two of the coins could be touching?

7. (*9*) Let G be a group of finite order and identity e. Suppose that ϕ is an automorphism of G onto itself with the following properties: (1) $\phi(x) = x$ if and only if $x = e$; (2) $\phi(\phi(x)) = x$ for each element x of G.

  (a) Give an example of a group and automorphism for which these conditions are satisfied.

  (b) Prove that G is commutative (i.e., $xy = yx$ for each pair x, y of elements in G).

8. (*11*) Let $b \geq 2$ be an integer base of numeration and let $1 \leq r \leq b-1$. Determine the sum of all r−digit numbers of the form

$$
\overline{a_{r-1}a_{r-2}\cdots a_2 a_1 a_0} \equiv a_{r-1}b^{r-1} + a_{r-2}b^{r-2} + \cdots + a_1 r + a_0
$$

whose digits increase strictly from left to right: $1 \leq a_{r-1} < a_{r-2} < \cdots < a_1 < a_0 \leq b-1$.

9. (*2*) For each positive integer n, let

$$
S_n = \sum_{k=1}^{n} \frac{2^k}{k^2}.
$$

Prove that S_{n+1}/S_n is not a rational function of n.

10. (*8*) A point is chosen at random (with the uniform distribution) on each side of a unit square. What is the probability that the four points are the vertices of a quadrilateral with area exceeding $\frac{1}{2}$?

Sunday, March 8, 2009

1. (*5*) Determine the supremum and the infimum of

$$
\frac{(x-1)^{x-1}x^x}{(x-(1/2))^{2x-1}}
$$

for $x > 1$.

2. (*7*) Let n and k be integers with $n \geq 0$ and $k \geq 1$. Let $\mathbf{x}_0, \mathbf{x}_1, \ldots, \mathbf{x}_n$ be $n+1$ distinct points in $\mathbb{R}^k$ and let $y_0, y_1, \ldots, y_n$ be $n+1$ real numbers (not necessarily distinct). Prove that there exists a polynomial p of degree at most n in the coordinates of $\mathbf{x}$ with respect to the standard basis for which $p(\mathbf{x}_i) = y_i$ for $0 \leq i \leq n$.

3. (*11*) For each positive integer n, let $p(n)$ be the product of all positive integral divisors of n. Is it possible to find two distinct positive integers m and n for which $p(m) = p(n)$?

4. (*4*) Let $\{a_n\}$ be a real sequence for which

$$\sum_{n=1}^{\infty} \frac{a_n}{n}$$

converges. Prove that

$$\lim_{n \to \infty} \frac{a_1 + a_2 + \cdots + a_n}{n} = 0.$$

5. (*7*) Find a 3×3 matrix A with elements in $\mathbb{Z}_2$ for which $A^7 = I$ and $A \neq I$. (Here, I is the identity matrix and $\mathbb{Z}_2$ is the field of two elements 0 and 1 where addition and multiplication are defined modulo 2.)

6. (*11*) Determine all solutions in nonnegative integers (x, y, z, w) to the equation

$$2^x 3^y - 5^z 7^w = 1.$$

7. (*3*) Let $n \geq 2$. Minimize $a_1 + a_2 + \cdots + a_n$ subject to the constraints $0 \leq a_1 \leq a_2 \leq \cdots \leq a_n$ and $a_1 a_2 + a_2 a_3 + \cdots + a_{n-1} a_n + a_n a_1 = 1$. (When $n = 2$, the latter condition is $a_1 a_2 = 1$; when $n \geq 3$, the sum on the left has exactly n terms.)

8. (*7*) Let a, b, c be members of a real inner-product space $(V, \langle, \rangle)$ whose norm is given by $\|x\|^2 = \langle x, x \rangle$. (You may assume that V is $\mathbb{R}^n$ if you wish.) Prove that

$$\|a + b\| + \|b + c\| + \|c + a\| \leq \|a\| + \|b\| + \|c\| + \|a + b + c\|$$

for $a, b, c, \in V$.

9. (*11*) Let p be a prime congruent to 1 modulo 4. For each real number x, let $\{x\} = x - \lfloor x \rfloor$ denote the fractional part of x. Determine

$$\sum \left\{ \left\{ \frac{k^2}{p} \right\} : 1 \leq k \leq \frac{1}{2}(p-1) \right\}.$$

10. (*10*) Suppose that a path on an $m \times n$ grid consisting of the lattice points $\{(x, y) : 1 \leq x \leq m, 1 \leq y \leq n\}$ (x and y both integers) consisting of $mn - 1$ unit segments begins at the point $(1, 1)$, passes through each point of the grid exactly once, does not intersect itself and finishes at the point (m, n). Show that the path partitions the rectangle bounded by the lines $x = 1$, $x = m$, $y = 1$, $y = n$ into two

subsets of equal area, the first consisting of regions opening to the left or up, and the second consisting of regions opening to the right or down.

Sunday, March 7, 2010

1. (8) Let F_1 and F_2 be the foci of an ellipse and P be a point in the plane of the ellipse. Suppose that G_1 and G_2 are points on the ellipse for which PG_1 and PG_2 are tangents to the ellipse. Prove that $\angle F_1PG_1 = \angle F_2PG_2$.

2. (10) Let $u_0 = 1$, $u_1 = 2$ and $u_{n+1} = 2u_n + u_{n-1}$ for $n \geq 1$. Prove that, for every nonnegative integer n,

$$u_n = \sum \left\{ \frac{(i+j+k)!}{i!j!k!} : i, j, k \geq 0, i + j + 2k = n \right\}.$$

3. (7) Let $\mathbf{a}$ and $\mathbf{b}$, the latter nonzero, be vectors in $\mathbb{R}^3$. Determine the value of λ for which the vector equation

$$\mathbf{a} - (\mathbf{x} \times \mathbf{b}) = \lambda \mathbf{b}$$

is solvable, and then solve it.

4. (10) The plane is partitioned into n regions by three families of parallel lines. What is the least number of lines to ensure that $n \geq 2010$?

5. (2) Let m be a natural number, and let $c, a_1, a_2, \ldots, a_m$ be complex numbers for which $|a_i| = 1$ for $i = 1, 2, \ldots, m$. Suppose also that

$$\lim_{n \to \infty} \sum_{i=1}^{m} a_i^n = c.$$

Prove that $c = m$ and that $a_i = 1$ for $i = 1, 2, \ldots, m$.

6. (2) Let $f(x)$ be a quadratic polynomial. Prove that there exist quadratic polynomials $g(x)$ and $h(x)$ for which

$$f(x)f(x+1) = g(h(x)).$$

7. (6) Suppose that f is a continuous real-valued function defined on the closed interval $[0, 1]$ and that

$$\left(\int_0^1 xf(x)\, dx \right)^2 = \left(\int_0^1 f(x)\, dx \right) \left(\int_0^1 x^2 f(x)\, dx \right).$$

Prove that there is a point $c \in (0, 1)$ for which $f(c) = 0$.

8. (7) Let A be an invertible symmetric $n \times n$ matrix with entries $\{a_{ij}\}$ in $\mathbb{Z}_2$. Prove that there is an $n \times n$ matrix with entries in $\mathbb{Z}_2$ such that $A = M^t M$ only if $a_{ii} \neq 0$ for some i.

9. (*6*) Let f be a real-valued function defined on $\mathbb{R}$ with a continuous third derivative, let $S_0 = \{x : f(x) = 0\}$, and, for $k = 1, 2, 3$, $S_k = \{x : f^{(k)}(x) = 0\}$, where $f^{(k)}$ denotes the kth derivative of f. Suppose also that $\mathbb{R} = S_0 \cup S_1 \cup S_2 \cup S_3$. Must f be a polynomial of degree not exceeding 2?

10. (*6*) Prove that the set $\mathbb{Q}$ of rationals can be written as the union of countably many subsets of $\mathbb{Q}$ each of which is dense in the set $\mathbb{R}$ of real numbers.

Sunday, March 6, 2011

1. (*6*) Let S be a nonvoid set of real numbers with the property that, for each real number x, there is a unique real number $f(x)$ belonging to S that is farthest from x, i.e., for each y in S distinct from $f(x)$, $|x - f(x)| > |x - y|$. Prove that S must be a singleton.

2. (*3*) Let u and v be positive reals. Minimize the larger of the two values

$$2u + \frac{1}{v^2} \text{ and } 2v + \frac{1}{u^2}.$$

3. (*10*) Suppose that S is a set of n nonzero real numbers such that exactly p of them are positive and exactly q are negative. Determine all the pairs (n, p) such that exactly half of the threefold products abc of distinct elements a, b, c of S are positive.

4. (*4*) Let $\{b_n : n \geq 1\}$ be a sequence of positive real numbers such that

$$3b_{n+2} \geq b_{n+1} + 2b_n$$

for every positive integer n. Prove that either the sequence converges or that it diverges to infinity.

5. (*2*) Solve the system

$$x + xy + xyz = 12$$
$$y + yz + yzx = 21$$
$$z + zx + zxy = 30.$$

6. (*2*) [The problem posed on the competition is not available for publication, as it appeared as Enigma problem 1610 in the August 25, 2010 issue of the *New Scientist*. It concerned the determination of the final score of a three game badminton match where the score of each competitor was in arithmetic progression. The full statement of the problem is available on the internet.]

7. (*7*) Suppose that there are 2011 students in a school and that each student has a certain number of friends among his schoolmates. It is assumed that if A is a friend of B, then B is a friend of A, and also that there may exist certain pairs that are not friends. Prove that there is a nonvoid subset S of these students for which every student in the school has an even number of friends in S.

8. (*9*) The set of transpositions of the symmetric group S_5 on $\{1, 2, 3, 4, 5\}$ is

$$\{(12), (13), (14), (15), (23), (24), (25), (34), (35), (45)\}$$

where (ab) denotes the permutation that interchanges a and b and leaves every other element fixed. Determine a product of all transpositions, each occuring exactly once, that is equal to the identity permutation ϵ, which leaves every element fixed.

9. (*7*) Suppose that A and B are two square matrices of the same order for which the indicated inverses exist. Prove that

$$(A + AB^{-1}A)^{-1} + (A + B)^{-1} = A^{-1}.$$

10. (*2*) Suppose that p is an odd prime. Determine the number of subsets S contained in $\{1, 2, \ldots, 2p - 1, 2p\}$ for which (a) S has exactly p elements, and (b) the sum of the elements of S is a multiple of p.

Saturday, March 10, 2012

1. (*8*) An equilateral triangle of side length 1 can be covered by five equilateral triangles of side length u. Prove that it can be covered by four equilateral triangles of side length u. (A triangle is a closed convex set that contains its three sides along with its interior.)

2. (*11*) Suppose that f is a function defined on the set $\mathbb{Z}$ of integers that takes integer values and satisfies the condition that $f(b) - f(a)$ is a multiple of $b - a$ for every pair a, b, of integers. Suppose also that p is a polynomial with integer coefficients such that $p(n) = f(n)$ for infinitely many integers n. Prove that $p(x) = f(x)$ for every positive integer x.

3. (*2*) Given the real numbers a, b, c not all zero, determine the real solutions x, y, z, u, v, w for the system of equations:

$$x^2 + v^2 + w^2 = a^2$$
$$u^2 + y^2 + w^2 = b^2$$
$$u^2 + v^2 + z^2 = c^2$$
$$u(y + z) + vw = bc$$
$$v(x + z) + wu = ca$$
$$w(x + y) + uv = ab.$$

4. (*11*) (a) Let n and k be positive integers. Prove that the least common multiple of $\{n, n+1, n+2, \ldots, n+k\}$ is equal to

$$rn\binom{n+k}{k}$$

for some positive integer r.

(b) For each positive integer k, prove that there exist infinitely many positive integers n, for which the number r defined in part (a) is equal to 1.

5. (*8*) Let $\mathfrak{C}$ be a circle and Q a point in the plane. Determine the locus of the centres of those circles that are tangent to $\mathfrak{C}$ and whose circumference passes through Q.

6. (*6*) Find all continuous real-valued functions defined on $\mathbb{R}$ that satisfy $f(0) = 0$ and

$$f(x) - f(y) = (x - y)g(x + y)$$

for some real-valued function $g(x)$.

7. (*6*) Consider the following problem:

Suppose that $f(x)$ is a continuous real-valued function defined on the interval $[0, 2]$ for which

$$\int_0^2 f(x)\, dx = \int_0^2 (f(x))^2\, dx.$$

Prove that there exists a number $c \in [0, 2]$ for which either $f(c) = 0$ or $f(c) = 1$.

(a) Criticize the following solution:

Solution. Clearly $\int_0^2 f(x)dx \geq 0$. By the Extreme Value Theorem, there exist numbers u and v in $[0, 2]$ for which $f(u) \leq f(x) \leq f(v)$ for $0 \leq x \leq 2$. Hence

$$f(u)\int_0^2 f(x)\, dx \leq \int_0^2 f(x)^2\, dx \leq f(v)\int_0^2 f(x)\, dx.$$

Since $\int_0^2 f(x)^2\, dx = 1 \cdot \int_0^2 f(x)\, dx$, by the Intermediate Value Theorem, there exists a number $c \in [0, 2]$ for which $f(c) = 1$. $\square$

(b) Show that there is a nontrivial function f that satisfies the conditions of the problem but that never assumes the value 1.

(c) Provide a complete solution of the problem.

8. (*8*) Determine the area of the set of points (x, y) in the plane that satisfy the two inequalities:

$$x^2 + y^2 \leq 2$$
$$x^4 + x^3 y^3 \leq xy + y^4.$$

9. (*10*) In a round-robin tournament of $n \geq 2$ teams, each pair of teams plays exactly one game that results in a win for one team and a loss for the other (there are no ties).
   (a) Prove that the teams can be labelled $t_1, t_2, \ldots, t_n$, so that, for each i with $1 \leq i \leq n - 1$, team t_i beats t_{i+1}.
   (b) Suppose that a team t has the property that, for each other team u, one can find a chain $u_1, u_2, \ldots, u_m$ of (possibly zero) distinct teams for which t beats u_1, u_i beats u_{i+1} for $1 \leq i \leq m - 1$ and u_m beats u. Prove that *all* of the n teams can be ordered as in (a) so that $t = t_1$ and each t_i beats t_{i+1} for $1 \leq i \leq n - 1$.
   (c) Let T denote the set of teams that can be labelled as t_1 in an ordering of teams as in (a). Prove that, in any ordering of teams as in (a), all the teams in T occur before all the teams that are not in T.

10. (*7*) Let A be a square matrix whose entries are complex numbers. Prove that $A^* = A$ if and only if $AA^* = A^2$.

Saturday, March 9, 2013

1. (*11*) (a) Let a be an odd positive integer exceeding 3, and let n be a positive integer. Prove that

$$a^{2^n} - 1$$

has at least $n + 1$ distinct prime divisors.

(b) When $a = 3$, determine all the positive integers n for which the assertion in (a) is false.

2. (*8*) $ABCD$ is a square; points U and V are situated on the respective sides BC and CD. Prove that the perimeter of triangle CUV is equal to twice the sidelength of the square if and only if $\angle UAV = 45°$.

3. (*6*) Let $f(x)$ be a convex increasing real-valued function defined on the closed interval $[0, 1]$ for which $f(0) = 0$ and $f(1) = 1$. Suppose that $0 < a < 1$ and that $b = f(a)$.
   (a) Prove that f is continuous on $(0, 1)$.
   (b) Prove that

$$0 \leq a - b \leq 2 \int_0^1 (x - f(x))\, dx \leq 1 - 4b(1 - a).$$

4. (*11*) Let S be the set of integers of the form $x^2 + xy + y^2$, where x and y are integers.
   (a) Prove that any prime p in S is either equal to 3 or is congruent to 1 modulo 6.
   (b) Prove that S includes all squares.
   (c) Prove that S is closed under multiplication.

5. (8) A point on an ellipse is joined to the ends of its major axis. Prove that the portion of a directrix intercepted by the two joining lines subtends a right angle at the corresponding focus.

6. (2) Let $p(x) = x^4 + ax^3 + bx^2 + cx + d$ be a polynomial with rational coefficients. Suppose that $p(x)$ has exactly one real root r. Prove that r is rational.

7. (7) Let $(V, \langle \cdot \rangle)$ be a two-dimensional inner product space over the complex field $\mathbf{C}$ and let z_1 and z_2 be unit vectors in V. Prove that

$$\sup\{|\langle z, z_1 \rangle \langle z, z_2 \rangle| : \|z\| = 1\} \geq \frac{1}{2}$$

with equality if and only if $\langle z_1, z_2 \rangle = 0$.

8. (7) For any real square matrix A, the adjugate matrix, adj A, has as its elements the cofactors of the transpose of A, so that

$$A \cdot \text{adj } A = \text{adj } A \cdot A = (\det A)I.$$

(a) Suppose that A is an invertible square matrix. Show that

$$(\text{adj } (A^t))^{-1} = (\text{adj } (A^{-1}))^t.$$

(b) Suppose that adj (A^t) is orthogonal (i.e., its inverse is its transpose). Prove that A is invertible.

(c) Let A be an invertible $n \times n$ square matrix and let $\det (tI - A) = t^n + c_1 t^{n-1} + \cdots + c_{n-1} t + c_n$ be the characteristic polynomial of the matrix A. Determine the characteristic polynomial of adj A.

9. (9) Let S be a set upon whose elements there is a binary operation $(x, y) \to xy$ which is associative (i.e. $x(yz) = (xy)z$). Suppose that there exists an element $e \in S$ for which $e^2 = e$ and that for each $a \in S$, there is at least one element b for which $ba = e$ and at most one element c for which $ac = e$. Prove that S is a group with this binary operation.

10. (6) (a) Let f be a real-valued function defined on the real number field $\mathbb{R}$ for which $|f(x) - f(y)| < |x - y|$ for any pair (x, y) of distinct elements of $\mathbb{R}$. Let $f^{(n)}$ denote the nth composite of f defined by $f^{(1)}(x) = f(x)$ and $f^{(n+1)}(x) = f(f^{(n)}(x))$ for $n \geq 2$. Prove that exactly one of the following situations must occur:
   (i) $\lim_{n \to +\infty} f^{(n)}(x) = +\infty$ for each real x;
   (ii) $\lim_{n \to +\infty} f^{(n)}(x) = -\infty$ for each real x;
   (iii) there is a real number z such that

$$\lim_{n \to +\infty} f^{(n)}(x) = z$$

   for each real x.
   (b) Give examples to show that each of the three cases in (a) can occur.

Sunday, March 9, 2014

1. (7) The *permanent*, per A, of an $n \times n$ matrix $A = (a_{i,j})$, is equal to the sum of all possible products of the form $a_{1,\sigma(1)} a_{2,\sigma(2)} \cdots a_{n,\sigma(n)}$, where σ runs over all the permutations on the set $\{1, 2, \ldots, n\}$. (This is similar to the definition of determinant, but there is no sign factor.) Show that, for any $n \times n$ matrix $A = (a_{i,j})$ with positive real terms,

$$\text{per } A \geq n! \left(\prod_{1 \leq i, j \leq n} a_{i,j} \right)^{\frac{1}{n}}.$$

2. (11) For a positive integer N written in base 10 numeration, N' denotes the integer with the digits of N written in reverse order. There are pairs of integers (A, B) for which A, A', B, B' are all distinct and $A \times B = B' \times A'$. For example,

$$3516 \times 8274 = 4728 \times 6153.$$

   (a) Determine a pair (A, B) as described above for which both A and B have two digits, and all four digits involved are distinct.
   (b) Are there any pairs (A, B) as described above for which A has two and B has three digits?

3. (11) Let n be a positive integer. A finite sequence $\{a_1, a_2, \ldots, a_n\}$ of positive integers a_i is said to be *tight* if and only if $1 \leq a_1 < a_2 < \cdots < a_n$, all $\binom{n}{2}$ differences $a_j - a_i$ with $i < j$ are distinct, and a_n is as small as possible.
   (a) Determine a tight sequence for $n = 5$.
   (b) Prove that there is a polynomial $p(n)$ of degree not exceeding 3 such that $a_n \leq p(n)$ for every tight sequence $\{a_i\}$ with n entries.

4. (6) Let $f(x)$ be a continuous real-valued function on $[0, 1]$ for which

$$\int_0^1 f(x)\, dx = 0 \qquad \text{and} \qquad \int_0^1 x\, f(x) dx = 1.$$

   (a) Give an example of such a function.
   (b) Prove that there is a nontrivial open interval I contained in $(0, 1)$ for which $|f(x)| > 4$ for $x \in I$.

5. (4) Let n be a positive integer. Prove that

$$\sum_{k=1}^{n} \frac{1}{k\binom{n}{k}} = \sum_{k=1}^{n} \frac{1}{k 2^{n-k}} = \frac{1}{2^{n-1}} \sum_{k=1}^{n} \frac{2^{k-1}}{k}$$

$$= \frac{1}{2^{n-1}} \sum \left\{ \binom{n}{k} : k \text{ odd}, 1 \leq k \leq n \right\}.$$

6. (2) Let $f(x) = x^6 - x^4 + 2x^3 - x^2 + 1$.
  (a) Prove that $f(x)$ has no positive real roots.
  (b) Determine a nonzero polynomial $g(x)$ of minimum degree for which all the coefficients of $f(x)g(x)$ are nonnegative rational numbers.
  (c) Determine a polynomial $h(x)$ of minimum degree for which all the coefficients of $f(x)h(x)$ are positive rational numbers.

7. (3) Suppose that $x_0, x_1, \ldots, x_n$ are real numbers. For $0 \le i \le n$, define
$$y_i = \max(x_0, x_1, \ldots, x_i).$$
Prove that
$$y_n^2 \le 4x_n^2 - 4\sum_{i=0}^{n-1} y_i(x_{i+1} - x_i).$$
When does equality occur?

8. (8) The hyperbola with equation $x^2 - y^2 = 1$ has two branches, as does the hyperbola with equation $y^2 - x^2 = 1$. Choose one point from each of the four branches of the locus of $(x^2 - y^2)^2 = 1$ such that area of the quadrilateral with these four vertices is minimized.

9. (4) Let $\{a_n\}$ and $\{b_n\}$ be positive real sequences such that
$$\lim_{n \to \infty} \frac{a_n}{n} = u > 0$$
and
$$\lim_{n \to \infty} \left(\frac{b_n}{a_n}\right)^n = v > 0.$$
Prove that
$$\lim_{n \to \infty} \left(\frac{b_n}{a_n}\right) = 1$$
and
$$\lim_{n \to \infty} (b_n - a_n) = u \log v.$$

10. (6) Does there exist a continuous real-valued function defined on $\mathbf{R}$ for which $f(f(x)) = -x$ for all $x \in \mathbb{R}$?

Sunday, March 8, 2015

1. (6) Suppose that u and v are two real-valued functions defined on the set of reals. Let $f(x) = u(v(x))$ and $g(x) = u(-v(x))$ for each real x. If $f(x)$ is continuous, must $g(x)$ also be continuous?

2. (8) Given $2n$ distinct points in space, the sum S of the lengths of all the segments joining pairs of them is calculated. Then n of the points are removed along with all the segments having at least one endpoint from among them. Prove that the sum of the lengths of all the remaining segments is less that $\frac{1}{2}S$.

3. *(6)* Let $f : [0, 1] \longrightarrow \mathbb{R}$ be continuously differentiable. Prove that

$$\left| \frac{f(0) + f(1)}{2} - \int_0^1 f(x)\,dx \right| \leq \frac{1}{4} \sup\{|f'(x)| : 0 \leq x \leq 1\}.$$

4. *(3)* Determine all the values of the positive integer $n \geq 2$ for which the following statement is true, and for each, indicate when equality holds.

For any nonnegative real numbers $x_1, x_2, \ldots, x_n$,

$$(x_1 + x_2 + \cdots + x_n)^2 \geq n(x_1 x_2 + x_2 x_3 + \cdots + x_{n-1} x_n + x_n x_1),$$

where the right side has n summands.

5. *(5)* Let $f(x)$ be a real polynomial of degree 4 whose graph has two real inflection points. There are three regions bounded by the graph and the line passing through these inflection points. Prove that two of these regions have equal area and that the area of the third region is equal to the sum of the other two areas.

6. *(11)* Using the digits 1, 2, 3, 4, 5, 6, 7, 8, each exactly once, create two numbers and form their product. For example, $472 \times 83{,}156 = 39{,}249{,}632$. What are the smallest and the largest values such a product can have?

7. *(5)* Determine

$$\int_0^2 \frac{e^x\,dx}{e^{1-x} + e^{x-1}}.$$

8. *(4)* Let $\{a_n\}$ and $\{b_n\}$ be two *decreasing* positive real sequences for which

$$\sum_{n=1}^{\infty} a_n = \infty$$

and

$$\sum_{n=1}^{\infty} b_n = \infty.$$

Let I be a subset of the natural numbers, and define the sequence $\{c_n\}$ by

$$c_n = \begin{cases} a_n, & \text{if } n \in I; \\ b_n, & \text{if } n \notin I. \end{cases}$$

Is it possible for $\sum_{n=1}^{\infty} c_n$ to converge?

9. *(7)* What is the dimension of the vector subspace of $\mathbf{R}^n$ generated by the set of vectors

$$(\sigma(1), \sigma(2), \sigma(3), \ldots, \sigma(n))$$

where σ runs through all $n!$ of the permutations of the first n natural numbers?

10. (2) (a) Let

$$g(x,y) = x^2y + xy^2 + xy + x + y + 1.$$

We form a sequence $\{x_0\}$ as follows: $x_0 = 0$. The next term x_1 is the unique solution -1 of the linear equation $g(t,0) = 0$. For each $n \geq 2$, x_n is the solution other than x_{n-2} of the equation $g(t, x_{n-1}) = 0$.

Let $\{f_n\}$ be the Fibonacci sequence determined by $f_0 = 0$, $f_1 = 1$ and $f_n = f_{n-1} + f_{n-2}$ for $n \geq 2$. Prove that, for any nonnegative integer k,

$$x_{2k} = \frac{f_k}{f_{k+1}} \quad \text{and} \quad x_{2k+1} = -\frac{f_{k+2}}{f_{k+1}}.$$

(b) Let

$$h(x,y) = x^2y + xy^2 + \beta xy + \gamma(x+y) + \delta$$

be a polynomial with real coefficients β, γ, δ. We form a bilateral sequence $\{x_n : n \in \mathbf{Z}\}$ as follows. Let $x_0 \neq 0$ be given arbitrarily. We select x_{-1} and x_1 to be the two solutions of the quadratic equation $h(t, x_0) = 0$ in either order. From here, we can define inductively the terms of the sequence for positive and negative values of the index so that x_{n-1} and x_{n+1} are the two solutions of the equation $h(t, x_n) = 0$. We suppose that at each stage, neither of these solutions is zero.

Prove that the sequence $\{x_n\}$ has period 5 (i.e. $x_{n+5} = x_n$ for each index n) if and only if $\gamma^3 + \delta^2 - \beta\gamma\delta = 0$.

CHAPTER 2

Algebra

2001:5. Let n be a positive integer and x a real number not equal to a nonnegative integer. Prove that

$$\frac{n}{x} + \frac{n(n-1)}{x(x-1)} + \frac{n(n-1)(n-2)}{x(x-1)(x-2)} + \cdots + \frac{n(n-1)(n-2)\cdots 1}{x(x-1)(x-2)\cdots(x-n+1)}$$
$$= \frac{n}{x-n+1}.$$

Solution 1. The result holds for $n = 1$ and $x \neq 0$. Suppose, as an induction hypothesis, the result holds for $n = k$ and x equal to any real number that is not a nonnegative integer; note that in this case, the left side has k terms. When $n = k + 1$,

$$\frac{k+1}{x} + \frac{(k+1)k}{x(x-1)} + \frac{(k+1)k(k-1)}{x(x-1)(x-2)} + \cdots + \frac{(k+1)k(k-1)\cdots 1}{x(x-1)(x-2)\cdots(x-k)}$$
$$= \frac{k+1}{x} + \frac{k+1}{x}\left[\frac{k}{x-1} + \frac{k(k-1)}{(x-1)(x-2)} + \cdots + \frac{k(k-1)\cdots 1}{(x-1)\cdots(x-k)}\right]$$
$$= \frac{k+1}{x} + \frac{k+1}{x}\left[\frac{k}{x-1-k+1}\right] = \frac{k+1}{x}\left[1 + \frac{k}{x-k}\right]$$
$$= \frac{k+1}{x}\left[\frac{x}{x-k}\right] = \frac{k+1}{x-(k+1)+1}$$

as desired. The result follows by induction.

Solution 2. [P. Gyrya] For real a and positive integer m, let $\binom{a}{m}$ denote $a(a-1)\cdots(a-m+1)/m!$ and let $\binom{a}{0}$ be 1. It is straightforward to establish that, for each positive integer k:

$$\binom{x}{k} = \binom{x-1}{k} + \binom{x-1}{k-1}$$

© Springer International Publishing Switzerland 2016
E.J. Barbeau, *University of Toronto Mathematics Competition*
(2001–2015), Problem Books in Mathematics,
DOI 10.1007/978-3-319-28106-3_2

and
$$\binom{x}{k} = \sum_{i=0}^{k} \binom{x-i-1}{k-i}.$$

The left side of the required equality is equal to

$$\frac{n!}{x\cdots(x-n+1)}\left[\binom{x-1}{n-1} + \binom{x-2}{n-2} + \cdots + \binom{x-n}{n-n}\right]$$

$$= \frac{n!}{x\cdots(x-n+1)}\binom{x}{n-1} = \frac{n}{x-n+1}$$

as desired.

2001:8 A regular heptagon (polygon with seven equal sides and seven equal angles) has diagonals of two different lengths. Let a be the length of a side, b be the length of a shorter diagonal and c be the length of a longer diagonal of a regular heptagon (so that $a < b < c$). Prove ONE of the following relationships:

$$\frac{a^2}{b^2} + \frac{b^2}{c^2} + \frac{c^2}{a^2} = 6$$

or

$$\frac{b^2}{a^2} + \frac{c^2}{b^2} + \frac{a^2}{c^2} = 5.$$

Solution 1. Let A, B, C, D, E be consecutive vertices of the regular heptagon. Let AB, AC and AD have respective lengths a, b, c, and let $\angle BAC = \theta$. Then $\theta = \pi/7$, the length of BC, of CD and of DE is a, the length of AE is c, $\angle CAD = \angle DAE = \theta$, since the angles are subtended by equal chords of the circumcircle of the heptagon, $\angle ADC = 2\theta$, $\angle ADE = \angle AED = 3\theta$ and $\angle ACD = 4\theta$. Triangles ABC and ACD can be glued together along BC and DC (with C on C) to form a triangle similar to $\triangle ABC$, whence

(2.1) $$\frac{a+c}{b} = \frac{b}{a}.$$

Triangles ACD and ADE can be glued together along CD and ED (with D on D) to form a triangle similar to triangle ABC, whence

(2.2) $$\frac{b+c}{c} = \frac{b}{a}.$$

Equation (2.2) can be rewritten as $\frac{1}{b} = \frac{1}{a} - \frac{1}{c}$. whence

$$b = \frac{ac}{c-a}.$$

Substituting this into (2.1) yields

$$\frac{(c+a)(c-a)}{ac} = \frac{c}{c-a}$$

which simplifies to

$$(2.3) \qquad\qquad a^3 - a^2c - 2ac^2 + c^3 = 0.$$

Note also from (2.1) that $b^2 = a^2 + ac$.

$$\frac{a^2}{b^2} + \frac{b^2}{c^2} + \frac{c^2}{a^2} - 6 = \frac{a^4c^2 + b^4a^2 + c^4b^2 - 6a^2b^2c^2}{a^2b^2c^2}$$

$$= \frac{a^4c^2 + (a^4 + 2a^3c + a^2c^2)a^2 + c^4(a^2 + ac) - 6a^2c^2(a^2 + ac)}{a^2b^2c^2}$$

$$= \frac{a^6 + 2a^5c - 4a^4c^2 - 6a^3c^3 + a^2c^4 + ac^5}{a^2b^2c^2}$$

$$= \frac{a(a^2 + 3ac + c^2)(a^3 - a^2c - 2ac^2 + c^3)}{a^2b^2c^2} = 0.$$

$$\frac{b^2}{a^2} + \frac{c^2}{b^2} + \frac{a^2}{c^2} - 5$$

$$= \frac{(a^4 + 2a^3c + a^2c^2)c^2 + a^2c^4 + a^4(a^2 + ac) - 5a^2c^2(a^2 + ac)}{a^2b^2c^2}$$

$$= \frac{a^6 + a^5c - 4a^4c^2 - 3a^3c^3 + 2a^2c^4}{a^2b^2c^2}$$

$$= \frac{a^2(a + 2c)(a^3 - a^2c - 2ac^2 + c^3)}{a^2b^2c^2} = 0.$$

Solution 2 (of the first result). [J. Chui] Let the heptagon be $ABCDEFG$ and $\theta = \pi/7$. Using the Law of Cosines in the indicated triangles ACD and ABC, we obtain the following:

$$\cos 2\theta = \frac{a^2 + c^2 - b^2}{2ac} = \frac{1}{2}\left(\frac{a}{c} + \frac{c}{a} - \frac{b^2}{ac}\right)$$

$$\cos 5\theta = \frac{2a^2 - b^2}{2a^2} = 1 - \frac{1}{2}\left(\frac{b}{a}\right)^2$$

from which, since $\cos 2\theta = -\cos 5\theta$,

$$-1 + \frac{1}{2}\left(\frac{b}{a}\right)^2 = \frac{1}{2}\left(\frac{a}{c} + \frac{c}{a} - \frac{b^2}{ac}\right)$$

or

$$(2.4) \qquad\qquad \frac{b^2}{a^2} = 2 + \frac{a}{c} + \frac{c}{a} - \frac{b^2}{ac}.$$

Examining triangles ABC and ADE, we find that $\cos\theta = b/2a$ and $\cos\theta = (2c^2 - a^2)/(2c^2) = 1 - (a^2/2c^2)$, so that

$$(2.5) \qquad\qquad \frac{a^2}{c^2} = 2 - \frac{b}{a}.$$

Examining triangles ADE and ACF, we find that $\cos 3\theta = a/2c$ and $\cos 3\theta = (2b^2 - c^2)/(2b^2)$, so that

$$(2.6) \qquad\qquad \frac{c^2}{b^2} = 2 - \frac{a}{c}.$$

Adding Eqs. (2.4), (2.5), (2.6) yields

$$\frac{b^2}{a^2} + \frac{c^2}{b^2} + \frac{a^2}{c^2} = 6 + \frac{c^2 - bc - b^2}{ac}.$$

By Ptolemy's Theorem, the sum of the products of pairs of opposite sides of a concylic quadrilaterial is equal to the product of the diagonals. Applying this to the quadrilaterals $ABDE$ and $ABCD$, respectively, yields $c^2 = a^2 + bc$ and $b^2 = ac + a^2$, whence $c^2 - bc - b^2 = a^2 + bc - bc - ac - a^2 = -ac$ and we find that

$$\frac{b^2}{a^2} + \frac{c^2}{b^2} + \frac{a^2}{c^2} = 6 - 1 = 5.$$

Solution 3. There is no loss of generality in assuming that the vertices of the heptagon are placed at the seventh roots of unity on the unit circle in the complex plane. Let $\zeta = \cos(2\pi/7) + i\sin(2\pi/7)$ be the fundamental seventh root of unity. Then $\zeta^7 = 1$, $1 + \zeta + \zeta^2 + \cdots + \zeta^6 = 0$ and (ζ, ζ^6), (ζ^2, ζ^5), (ζ^3, ζ^4) are pairs of complex conjugates. We have that

$$a = |\zeta - 1| = |\zeta^6 - 1|$$

$$b = |\zeta^2 - 1| = |\zeta^9 - 1|$$

$$c = |\zeta^3 - 1| = |\zeta^4 - 1|.$$

It follows from this that

$$\frac{b}{a} = |\zeta + 1| \qquad \frac{c}{b} = |\zeta^2 + 1| \qquad \frac{a}{c} = |\zeta^3 + 1|,$$

whence

$$\frac{b^2}{a^2} + \frac{c^2}{b^2} + \frac{a^2}{c^2} = (\zeta + 1)(\zeta^6 + 1) + (\zeta^2 + 1)(\zeta^5 + 1) + (\zeta^3 + 1)(\zeta^4 + 1)$$

$$= 2 + \zeta + \zeta^6 + 2 + \zeta^2 + \zeta^5 + 2 + \zeta^3 + \zeta^4$$

$$= 6 + (\zeta + \zeta^2 + \zeta^3 + \zeta^4 + \zeta^5 + \zeta^6) = 6 - 1 = 5.$$

Also

$$\frac{a}{b} = |\zeta^4 + \zeta^2 + 1| \qquad \frac{b}{c} = |\zeta^6 + \zeta^3 + 1| \qquad \frac{c}{a} = |\zeta^2 + \zeta + 1|,$$

whence

$$\frac{a^2}{b^2} + \frac{b^2}{c^2} + \frac{c^2}{a^2} = (\zeta^4 + \zeta^2 + 1)(\zeta^3 + \zeta^5 + 1) + (\zeta^6 + \zeta^3 + 1)(\zeta + \zeta^4 + 1)$$
$$+ (\zeta^2 + \zeta + 1)(\zeta^5 + \zeta^6 + 1)$$
$$= (3 + 2\zeta^2 + \zeta^3 + \zeta^4 + 2\zeta^5) + (3 + \zeta + 2\zeta^3 + 2\zeta^4 + \zeta^6)$$
$$+ (3 + 2\zeta + \zeta^2 + \zeta^5 + 2\zeta^6)$$
$$= 9 + 3(\zeta + \zeta^2 + \zeta^3 + \zeta^4 + \zeta^5 + \zeta^6) = 9 - 3 = 6.$$

Solution 4. Suppose that the circumradius of the heptagon is 1. By considering isosceles triangles with base equal to the sides or diagonals of the heptagon and apex at the centre of the circumcircle, we see that

$$a = 2\sin\theta = 2\sin 6\theta = -2\sin 8\theta$$
$$b = 2\sin 2\theta = -2\sin 9\theta$$
$$c = 2\sin 3\theta = 2\sin 4\theta$$

where $\theta = \pi/7$ is half the angle subtended at the circumcentre by each side of the heptagon. Observe that

$$\cos 2\theta = \frac{1}{2}(\zeta + \zeta^6) \qquad \cos 4\theta = \frac{1}{2}(\zeta^2 + \zeta^5) \qquad \cos 6\theta = \frac{1}{2}(\zeta^3 + \zeta^4)$$

where ζ is the fundamental primitive root of unity. We have that

$$\frac{b}{a} = 2\cos\theta = 2\cos 6\theta \qquad \frac{c}{b} = 2\cos 2\theta \qquad \frac{a}{c} = -2\cos 4\theta$$

whence

$$\frac{b^2}{a^2} + \frac{c^2}{b^2} + \frac{a^2}{c^2} = 4\cos^2 6\theta + 4\cos^2 2\theta + 4\cos^2 4\theta$$
$$= (\zeta^3 + \zeta^4)^2 + (\zeta + \zeta^6)^2 + (\zeta^2 + \zeta^5)^2$$
$$= \zeta^6 + 2 + \zeta + \zeta^2 + 2 + \zeta^5 + \zeta^4 + 2 + \zeta^3 = 6 - 1 = 5.$$

Also

$$\frac{a}{b} = \frac{\sin 6\theta}{\sin 2\theta} = 4\cos^2 2\theta - 1 = (\zeta + \zeta^6)^2 - 1 = 1 + \zeta^2 + \zeta^5$$
$$-\frac{b}{c} = \frac{\sin 9\theta}{\sin 3\theta} = 4\cos^2 3\theta - 1 = 4\cos^2 4\theta - 1$$
$$= (\zeta^2 + \zeta^5)^2 - 1 = 1 + \zeta^4 + \zeta^3$$
$$\frac{c}{a} = \frac{\sin 3\theta}{\sin\theta} = 4\cos^2 6\theta - 1 = (\zeta^3 + \zeta^4)^2 - 1 = 1 + \zeta^6 + \zeta,$$

whence

$$\frac{a^2}{b^2} + \frac{b^2}{c^2} + \frac{c^2}{a^2} = (3 + 2\zeta^2 + \zeta^3 + \zeta^4 + 2\zeta^5) + (3 + \zeta + 2\zeta^3 + 2\zeta^4 + \zeta^6)$$
$$+ (3 + 2\zeta + \zeta^2 + \zeta^5 + 2\zeta^6)$$
$$= 9 + 3(\zeta + \zeta^2 + \zeta^3 + \zeta^4 + \zeta^5 + \zeta^6) = 9 - 3 = 6.$$

Comment. The regular heptagon is a rich source of interesting relationships. See the paper *Golden fields; a case for the heptagon* by Peter Steinbach in *Mathematics Magazine* 70:1 (February, 1997), 22–31.

2002:2. Angus likes to go to the movies. On Monday, standing in line, he noted that the fraction x of the line was in front of him, while $1/n$ of the line was behind him. On Tuesday, the same fraction x of the line was in front of him, while $1/(n+1)$ of the line was behind him. On Wednesday, the same fraction x of the line was in front of him, while $1/(n+2)$ of the line was behind him. Determine a value of n for which this is possible.

Answer. When $x = 5/6$, he could have $1/7$ of a line of 42 behind him, $1/8$ of a line of 24 behind him and $1/9$ of a line of 18 behind him. When $x = 11/12$, he could have $1/14$ of a line of 84 behind him, $1/15$ of a line of 60 behind him and $1/16$ of a line of 48 behind him. When $x = 13/15$, he could have $1/8$ of a line of 120 behind him, $1/9$ of a line of 45 behind him and $1/10$ of a line of 30 behind him.

Solution. The strategy in this solution is to try to narrow down the search by considering a special case. Suppose that $x = (u-1)/u$ for some positive integer exceeding 1. Let $1/(u+p)$ be the fraction of the line behind Angus. Then Angus himself represents this fraction of the line:

$$1 - \left(\frac{u-1}{u} + \frac{1}{u+p} \right) = \frac{p}{u(u+p)},$$

so that there would be $u(u+p)/p$ people in line. To make this an integer, we can arrange that u is a multiple of p. To get an integer for $p = 1, 2, 3$, take u to be any multiple of 6. Thus, we can arrange that x is any of $5/6$, $11/12$, $17/18$, $23/24$, and so on.

Comment 1. The solution indicates how we can select x for which the amount of the line behind Angus is represented by any number of consecutive integer reciprocals. For example, in the case of $x = 11/12$, he could also have $1/13$ of a line of 156 behind him. Another strategy might be to look at $x = (u-2)/u$, i.e. successively at $x = 3/5, 5/7, 7/9, \cdots$. In this case, we assume that $1/(u-p)$ of the line is behind him, and need to ensure that $u - 2p$ is a positive divisor of $u(u-p)$ for three consecutive values of $u - p$. If u is odd, we can achieve this with u any odd multiple of 15 and with $p = \frac{1}{2}(u-1), \frac{1}{2}(u-3), \frac{1}{2}(u-5)$.

Comment 2. With the same fraction in front on 2 days, suppose that $1/n$ of a line of u people is behind Angus on the first day, and $1/(n+1)$ of a line of v people is behind him on the second day. Then

$$\frac{1}{u} + \frac{1}{n} = \frac{1}{v} + \frac{1}{n+1}$$

so that $uv = n(n+1)(u-v)$. This yields both $(n^2 + n - v)u = (n^2 + n)v$ and $(n^2 + n + u)v = (n^2 + n)u$, leading to

$$u - v = \frac{u^2}{n^2 + n + u} = \frac{v^2}{n^2 + n - v}.$$

Two immediate possibilities are $(n, u, v) = (n, n+1, n)$ and $(n, u, v) = (n, n(n+1), \frac{1}{2}n(n+1))$. To get some more, taking $u - v = k$, we get the quadratic equation

$$u^2 - ku - k(n^2 + n) = 0$$

with discriminant

$$\Delta = k^2 + 4(n^2 + n)k = [k + 2(n^2 + n)]^2 - 4(n^2 + n)^2,$$

a pythagorean relationship when Δ is square and the equation has integer solutions. Select α, β, γ so that $\gamma\alpha\beta = n^2 + n$ and let $k = \gamma(\alpha^2 + \beta^2 - 2\alpha\beta) = \gamma(\alpha - \beta)^2$; this will make the discriminant Δ equal to a square.

Taking $n = 3$, for example, yields the possibilities $(u, v) = (132, 11)$, $(60, 10)$, $(36, 9)$, $(24, 8)$, $(12, 6)$, $(6, 4)$, $(4, 3)$. In general, we find that $(n, u, v) = (n, \gamma\alpha(\alpha - \beta), \gamma\beta(\alpha - \beta))$ when $n^2 + n = \gamma\alpha\beta$ with $\alpha > \beta$. It turns out that $k = u - v = \gamma(\alpha - \beta)^2$.

2003:7. Suppose that the polynomial $f(x)$ of degree $n \geq 1$ has all real roots and that $\lambda > 0$. Prove that the set $\{x \in \mathbb{R} : |f(x)| \leq \lambda|f'(x)|\}$ is a finite union of closed intervals whose total length is equal to $2n\lambda$.

The solution to this problem appears in Chap. 6.

2004:1. Prove that, for any nonzero complex numbers z and w,

$$(|z| + |w|)\left|\frac{z}{|z|} + \frac{w}{|w|}\right| \leq 2|z + w|.$$

Solution 1.

$$(|z| + |w|)\left|\frac{z}{|z|} + \frac{w}{|w|}\right|$$

$$= \left|z + w + \frac{|z|w}{|w|} + \frac{|w|z}{|z|}\right|$$

$$\leq |z + w| + \frac{1}{|z||w|}|\bar{z}zw + \bar{w}zw|$$

$$= |z + w| + \frac{|zw|}{|z||w|}|\bar{z} + \bar{w}| = 2|z + w|.$$

Solution 2. Let $z = ae^{i\alpha}$ and $w = be^{i\beta}$, with a and b real and positive. Then the left side is equal to

$$|(a+b)(e^{i\alpha} + e^{i\beta})| = |ae^{i\alpha} + ae^{i\beta} + be^{i\alpha} + be^{i\beta}|$$
$$\leq |ae^{i\alpha} + be^{i\beta}| + |ae^{i\beta} + be^{i\alpha}|.$$

Observe that

$$|z + w|^2 = |(ae^{i\alpha} + be^{i\beta})(ae^{-i\alpha} + be^{-i\beta})|$$
$$= a^2 + b^2 + ab[e^{i(\alpha-\beta)} + e^{i(\beta-\alpha)}]$$
$$= |(ae^{i\beta} + be^{i\alpha})(ae^{-i\beta} + be^{-i\alpha})|$$

from which we find that the left side does not exceed

$$|ae^{i\alpha} + be^{i\beta}| + |ae^{i\beta} + be^{i\alpha}| = 2|ae^{i\alpha} + be^{i\beta}| = 2|z + w|.$$

Solution 3. Let $z = ae^{i\alpha}$ and $w = be^{i\beta}$, where a and b are positive reals. Then the inequality is equivalent to

$$\left| \frac{1}{2}(e^{i\alpha} + e^{i\beta}) \right| \leq |\lambda e^{i\alpha} + (1 - \lambda)e^{i\beta}|$$

where $\lambda = a/(a + b)$. But this simply says that the midpoint of the segment joining $e^{i\alpha}$ and $e^{i\beta}$ on the unit circle in the Argand diagram representing complex numbers as points in the plane is at least as close to the origin as another point on the segment.

Solution 4. [G. Goldstein] The inequality is equivalent to

$$(|t| + 1) \left| \frac{t}{|t|} + 1 \right| \leq 2|t + 1|$$

where $t = z/w$.

Let $t = r(\cos\theta + i\sin\theta)$. Then the inequality becomes

$$(r + 1)\sqrt{(\cos\theta + 1)^2 + \sin^2\theta} \leq 2\sqrt{(r\cos\theta + 1)^2 + r^2\sin^2\theta}$$
$$= 2\sqrt{r^2 + 2r\cos\theta + 1}.$$

Now,

$$4(r^2 + 2r\cos\theta + 1) - (r + 1)^2(2 + 2\cos\theta)$$
$$= 2r^2(1 - \cos\theta) + 4r(\cos\theta - 1) + 2(1 - \cos\theta)$$
$$= 2(r - 1)^2(1 - \cos\theta) \geq 0,$$

from which the inequality follows.

Solution 5. [R. Mong] Consider complex numbers as vectors in the plane. $q = (|z|/|w|)w$ is a vector of magnitude z in the direction w and $p = (|w|/|z|)z$ is a vector of magnitude w in the direction z. A reflection about the angle bisector of vectors z and w interchanges p and w, q and z. Hence $|p + q| = |w + z|$. Therefore

$$(|z| + |w|) \left| \frac{z}{|z|} + \frac{w}{|w|} \right|$$
$$= |z + q + p + w| \le |z + w| + |p + q|$$
$$= 2|z + w|.$$

2004:7. Let a be a parameter. Define the sequence $\{f_n(x) : n = 0, 1, 2, \cdots\}$ of polynomials by

$$f_0(x) \equiv 1$$

and

$$f_{n+1}(x) = x f_n(x) + f_n(ax)$$

for $n \ge 0$.

(a) Prove that, for all n, x,

$$f_n(x) = x^n f_n(1/x).$$

(b) Determine a formula for the coefficient of x^k ($0 \le k \le n$) in $f_n(x)$.

Solution. The polynomial $f_n(x)$ has degree n for each n, and we will write

$$f_n(x) = \sum_{k=0}^{n} b(n, k) x^k.$$

Then

$$x^n f_n(1/x) = \sum_{k=0}^{n} b(n, k) x^{n-k} = \sum_{k=0}^{n} b(n, n - k) x^k.$$

Thus, (a) is equivalent to $b(n, k) = b(n, n - k)$ for $0 \le k \le n$.

When $a = 1$, it can be established by induction that $f_n(x) = (x + 1)^n = \sum_{k=0}^{n} \binom{n}{k} x^n$. Also, when $a = 0$, $f_n(x) = x^n + x^{n-1} + \cdots + x + 1 = (x^{n+1} - 1)(x - 1)^{-1}$. Thus, (a) holds in these cases and $b(n, k)$ is respectively equal to $\binom{n}{k}$ and 1.

Suppose, henceforth, that $a \ne 1$. For $n \ge 0$,

$$f_{n+1}(x) = \sum_{k=0}^{n} b(n, k) x^{k+1} + \sum_{k=0}^{n} a^k b(n, k) x^k$$

$$= \sum_{k=1}^{n} b(n, k - 1) x^k + b(n, n) x^{n+1} + b(n, 0) + \sum_{k=1}^{n} a^k b(n, k) x^k$$

$$= b(n, 0) + \sum_{k=1}^{n} [b(n, k - 1) + a^k b(n, k)] x^k + b(n, n) x^{n+1},$$

whence $b(n+1,0) = b(n,0) = b(1,0)$ and $b(n+1,n+1) = b(n,n) = b(1,1)$ for all $n \geq 1$. Since $f_1(x) = x+1$, $b(n,0) = b(n,n) = 1$ for each n. Also

$$(2.7) \qquad b(n+1,k) = b(n,k-1) + a^k b(n,k)$$

for $1 \leq k \leq n$.

We conjecture what the coefficients $b(n,k)$ are from an examination of the first few terms of the sequence:

$$f_0(x) = 1; \quad f_1(x) = 1+x; \quad f_2(x) = 1+(a+1)x + x^2;$$

$$f_3(x) = 1+(a^2+a+1)x + (a^2+a+1)x^2 + x^3;$$

$$f_4(x) = 1+(a^3+a^2+a+1)x+(a^4+a^3+2a^2+a+1)x^2+(a^3+a^2+a+1)x^3+x^4;$$

$$f_5(x) = (1+x^5)+(a^4+a^3+a^2+a+1)(x+x^4)+(a^6+a^5+2a^4+2a^3+2a^2+a+1)$$
$$\times(x^2+x^3).$$

We make the empirical observation that

$$(2.8) \qquad b(n+1,k) = a^{n+1-k}b(n,k-1) + b(n,k)$$

which, with (2.7), yields

$$(a^{n+1-k} - 1)b(n,k-1) = (a^k - 1)b(n,k)$$

so that

$$b(n+1,k) = \left[\frac{a^k - 1}{a^{n+1-k} - 1} + a^k\right] b(n,k) = \left[\frac{a^{n+1} - 1}{a^{n+1-k} - 1}\right] b(n,k)$$

for $n \geq k$. This leads to the conjecture that, when $n > k$,

$$(2.9) \qquad b(n,k) = \left(\frac{(a^n - 1)(a^{n-1} - 1)\cdots(a^{k+1} - 1)}{(a^{n-k} - 1)(a^{n-k-1} - 1)\cdots(a - 1)}\right) b(k,k)$$

where $b(k,k) = 1$.

We establish this conjecture. Let $c(n,k)$ be the right side of (2.9) for $1 \leq k \leq n-1$ and $c(n,n) = 1$. Then $c(n,0) = b(n,0) = c(n,n) = b(n,n) = 1$ for each n. In particular, $c(n,k) = b(n,k)$ when $n = 1$.

We show that

$$c(n+1,k) = c(n,k-1) + a^k c(n,k)$$

for $1 \leq k \leq n$, which will, through an induction argument, imply that $b(n,k) = c(n,k)$ for $0 \leq k \leq n$. The right side is equal to

$$\left(\frac{a^n - 1}{a^{n-k} - 1}\right)\cdots\left(\frac{a^{k+1} - 1}{a - 1}\right)\left[\frac{a^k - 1}{a^{n-k+1} - 1} + a^k\right]$$

$$= \frac{(a^{n+1} - 1)(a^n - 1)\cdots(a^{k+1} - 1)}{(a^{n+1-k} - 1)(a^{n-k} - 1)\cdots(a - 1)} = c(n+1,k)$$

as desired. Thus, we now have a formula for $b(n,k)$ as required in (b).

Finally, (a) can be established in a straightforward way, either from the formula (2.9) or using the pair of recursions (2.7) and (2.8).

2005:5. Let $f(x)$ be a polynomial with real coefficients, evenly many of which are nonzero, which is *palindromic*. This means that the coefficients read the same in either direction, i.e. $a_k = a_{n-k}$ if $f(x) = \sum_{k=0}^{n} a_k x^k$, or, alternatively, $f(x) = x^n f(1/x)$, where n is the degree of the polynomial. Prove that $f(x)$ has at least one root of absolute value 1.

The solution of this problem appears in Chap. 6.

2005:7. Let $f(x)$ be a nonconstant polynomial that takes only integer values when x is an integer, and let P be the set of all primes that divide $f(m)$ for at least one integer m. Prove that P is an infinite set.

The solution to this problem appears in Chap. 11.

2006:3. Let $p(x)$ be a polynomial of positive degree n with n distinct real roots $a_1 < a_2 < \cdots < a_n$. Let b be a real number for which $2b < a_1 + a_2$. Prove that

$$2^{n-1}|p(b)| \geq |p'(a_1)(b - a_1)|.$$

Solution. Wolog, let $p(x)$ have leading coefficient 1. Observe that, for $i > 1$,

$$a_i - a_1 = a_i - b + b - a_1 < a_i - b + a_2 - b < 2(a_i - b)$$

and that $p(x) = \prod(x - a_i)$, from which $p'(a_1) = \prod_{i \geq 2}(a_1 - a_i)$. Then

$$|p(b)| = |b - a_1| \prod_{i \geq 2} |b - a_i|$$

$$\geq |b - a_1| \prod_{i \geq 2} \frac{1}{2}(a_i - a_1) = |b - a_1||p'(a_i)|2^{-(n-1)}.$$

2006:9. A high school student asked to solve the surd equation

$$\sqrt{3x - 2} - \sqrt{2x - 3} = 1$$

gave the following answer: *Squaring both sides leads to*

$$3x - 2 - 2x - 3 = 1$$

so $x = 6$. The answer is, in fact, correct.

Show that there are infinitely many real quadruples (a, b, c, d) for which this method leads to a correct solution of the surd equation

$$\sqrt{ax - b} - \sqrt{cx - d} = 1.$$

Solution 1. Solving the general equation properly leads to

$$\sqrt{ax - b} - \sqrt{cx - d} = 1 \implies ax - b = 1 + cx - d + 2\sqrt{cx - d}$$

$$\implies (a - c)x = (b + 1 - d) + 2\sqrt{cx - d}.$$

To make the manipulation simpler, specialize to $a = c + 1$ and $d = b + 1$. Then the equation becomes

$$x^2 = 4(cx - d) \implies 0 = x^2 - 4cx + 4d.$$

Using the student's "method" to solve the same equation gives $ax - b - cx - d = 1$ which yields $x = (1 + b + d)/(a - c) = 2d$. So, for the "method" to work, we need

$$0 = 4d^2 - 8cd + 4d = 4d(d - 2c + 1)$$

which can be achieved by making $2c = d + 1$. So we can take

$$(a, b, c, d) = (t + 1, 2(t - 1), t, 2t - 1)$$

for some real t. The original problem corresponds to $t = 2$.

The equation

$$\sqrt{(t + 1)x - 2(t - 1)} - \sqrt{tx - (2t - 1)} = 1$$

is satisfied by $x = 2$ and $x = 4t - 2$. The first solution works for all values of t, while the second is valid if and only if $t \geq \frac{1}{2}$. The equation $(t + 1)x - 2(t - 1) - tx - (2t - 1) = 1$ is equivalent to $x = 4t - 2$.

Solution 2. [G. Goldstein] *Analysis.* We want to solve simultaneously the equations

(2.10) $\sqrt{ax - b} - \sqrt{cx - d} = 1$

and

(2.11) $ax - b - cx - d = 1.$

From (2.10), we find that

(2.12) $ax - b = 1 + (cx - d) + 2\sqrt{cx - d}.$

From (2.11) and (2.12), we obtain that $d = \sqrt{cx - d}$, so that $x = (d^2 + d)/c$. From (2.11), we have that $x = (1 + b + d)/(a - c)$.

Select a, c, d so that $d > 0$ and $ac(a - c) \neq 0$, and choose b to satisfy

$$\frac{d^2 + d}{c} = \frac{1 + b + d}{a - c}.$$

Let

$$x = \frac{d^2 + d}{c} = \frac{1 + b + d}{a - c} = \frac{(d^2 + d) + (1 + b + d)}{c + (a - c)} = \frac{(d + 1)^2 + b}{a}.$$

Then

$$\sqrt{ax - b} - \sqrt{cx - d} = \sqrt{(d + 1)^2} - \sqrt{d^2} = (d + 1) - d = 1$$

and

$$ax - b - cx - d = (d + 1)^2 + b - b - d^2 - d - d = 1.$$

Comments. In Solution 2, if we take $c = d = 1$, we get the family of parameters $(a, b, c, d) = (a, 2a - 4, 1, 1)$. R. Barrington Leigh found the set of parameters given in Solution 1. A. Feizmohammadi provided the parameters $(a, b, c, d) = (2c, 0, c, 1)$.

2008:4. Suppose that u, v, w, z are complex numbers for which $u + v + w + z = u^2 + v^2 + w^2 + z^2 = 0$. Prove that

$$(u^4 + v^4 + w^4 + z^4)^2 = 4(u^8 + v^8 + w^8 + z^8).$$

Solution. Let u, v, w, z be the roots of a monic quartic polynomial $f(t)$. Since $u + v + w + z = 0$ and

$$uv + uw + uz + vw + vz + wz = \frac{1}{2}[(u + v + w + z)^2 - (u^2 + v^2 + w^2 + z^2)] = 0,$$

we must have that $f(t) = t^4 - at - b$ for some complex numbers a and b.

Suppose that $s_n = u^n + v^n + w^n + z^n$ for positive integers n. Then $s_4 = 4b$ and $s_{k+4} = as_{k+1} + bs_k$ for $k \geq 1$. Therefore, $s_5 = 0$ and $s_8 = as_5 + bs_4 = 4b^2$, so that $4s_8 = (4b)^2 = s_4^2$, as desired.

2008:9. For each positive integer n, let

$$S_n = \sum_{k=1}^{n} \frac{2^k}{k^2}.$$

Prove that S_{n+1}/S_n is not a rational function of n.

Solution 1. Assume that S_{n+1}/S_n is a rational function of n. Then

$$\frac{S_n}{2^n n^{-2}} = \frac{S_n}{S_n - S_{n-1}} = \frac{1}{1 - \frac{S_{n-1}}{S_n}}$$

is a rational function. From this, it follows that $S_n = r(n)2^n n^{-2}$ for some rational function r. Since

$$\frac{2^{n+1}}{(n+1)^2} = S_{n+1} - S_n = r(n+1)\frac{2^{n+1}}{(n+1)^2} - r(n)\frac{2^n}{n^2},$$

we have that

$$2n^2 r(n+1) - (n+1)^2 r(n) = 2n^2.$$

Since this equation holds for infinitely many values of n, we have the corresponding rational function identity. Letting $r(x) = f(x)/g(x)$, where f and g are coprime polynomials, we obtain the equation

$$2x^2 f(x+1)g(x) - (x+1)^2 f(x)g(x+1) = 2x^2 g(x)g(x+1).$$

The polynomial $g(x)$ cannot be a constant (look at the leading coefficients).

There exists an integer n, namely 0, for which the polynomials $g(x)$ and $g(x+n)$ have a common divisor of positive degree. However, if n exceeds the maximum absolute value of a root of $g(x)$, then $g(x)$ and $g(x+n)$ are coprime. Let N be the largest integer for which $g(x)$ and $g(x + N)$ have a common irreducible divisor $u(x)$ of positive degree. Then $u(x - N)$ divides $g(x)$ and so divides $(x+1)^2 f(x)g(x+1)$. Since f and g are coprime, $u(x - N)$ divides $(x + 1)^2 g(x + 1)$. Suppose that $u(x - N)$ divides $g(x + 1)$; then $u(x)$ must divide $g(x + N + 1)$, which contradicts the determination of N. Therefore, $u(x - N)$ divides $(x + 1)^2$, and so $u(x + 1)$ divides $(x + N + 2)^2$.

Since $u(x+1)$ divides $g(x+1)$, $u(x+1)$ must divide $2x^2 f(x+1)g(x)$. Since $u(x+1)$ does not divide $f(x+1)$, $u(x+1)$ divides $2x^2 g(x)$. Since $u(x+1)$ divides $g(x+N+1)$, $u(x+1)$ cannot divide $g(x)$. Therefore, $u(x+1)$ must divide x^2, as well as $(x+N+2)^2$. But this is an impossibility. This contradiction yields the result of the problem.

Solution 2. [C. Ochanine] Since

$$\frac{S_{n+1}}{S_n} - 1 = \frac{S_{n+1} - S_n}{S_n} = \frac{2^{n+1}}{(n+1)^2 S_n},$$

it is enough to show that $S_n/2^n$ is not a rational function. Let $f(n) = S_n/2^n$. Then

$$f(n+1) = \frac{1}{2}f(n) + \frac{1}{(n+1)^2}.$$

for every positive integer n.

If f were rational, then we would have

$$f(x+1) = \frac{1}{2}f(x) + \frac{1}{(x+1)^2}$$

for all real x. Since, by the definition, $f(1)$ is finite, so also is $f(0)$. Substituting $x = -1$ in the foregoing equation, we see that -1 is a pole of $f(x)$. Since

$$f(x+2) = \frac{1}{2}f(x+1) + \frac{1}{(x+2)^2} = \frac{1}{4}f(x) + \frac{1}{2(x+1)^2} + \frac{1}{(x+2)^2},$$

we see that -2 is also a pole of $f(x)$. Continuing on, we find that every negative integer is a pole of $f(x)$, contradicting its rationality.

The desired result follows.

[Problem **2008:9** was contributed by Franklin Vera Pacheco.]

2009:2. Let n and k be integers with $n \geq 0$ and $k \geq 1$. Let $\mathbf{x}_0, \mathbf{x}_1, \ldots, \mathbf{x}_n$ be $n+1$ distinct points in $\mathbb{R}^k$ and let $y_0, y_1, \ldots, y_n$ be $n+1$ real numbers (not necessarily distinct). Prove that there exists a polynomial p of degree at most n in the coordinates of $\mathbf{x}$ for which $p(\mathbf{x}_i) = y_i$ for $0 \leq i \leq n$.

The solution to this problem appears in Chap. 7.

2010:5. Let m be a natural number, and let $c, a_1, a_2, \cdots, a_m$ be complex numbers for which $|a_i| = 1$ for $i = 1, 2, \cdots, m$. Suppose also that

$$\lim_{n \to \infty} \sum_{i=1}^{m} a_i^n = c.$$

Prove that $c = m$ and that $a_i = 1$ for $i = 1, 2, \cdots, m$.

Solution. If $a_i = e^{i\alpha_i}$, then either the sequence $\{a_i^n\}$ is periodic and assumes the value 1 infinitely often (when α_i is a rational multiple of π) or has a subsequence whose limit is 1 (when α_i is not a rational multiple of π).

[In the latter case, we can find an increasing subsequence $\{n_k\}$ of natural numbers for which $a_i^{n_k}$ converges, so that $a_i^{n_{k+1}-n_k}$ converges to 1.]

We prove that there is a subsequence $\{n_k\}$ of natural numbers for which $\lim_k a_i^{n_k} = 1$ for each $1 \le i \le m$. Proceed by induction on m. When $m = 1$, the limit of any subsequence of $\{a_1^n\}$ is equal to the limit of the whole sequence, so that $c = 1$ in this case. In fact, we can go further: $a_1 = \lim a_1^{n+1} = \lim a_1^n = 1$.

Suppose the induction hypothesis holds for $m - 1$. Then there is a subsequence S_1 of natural numbers such that $\{a_i^n\}$ has limit 1 along this subsequence for $1 \le i \le m - 1$. We can find a subsequence S_2 along which the sequence $\{a_m^n\}$ converges to some limit b on the unit circle. Let S_3 be the sequence of differences of consecutive terms in the sequence S_2 (so that the sequence along S_3 consists of quotients of consecutive terms of the sequences along S_2). Then, for $1 \le i \le m$, the sequence $\{a_i^n\}$ converges to 1 along S_3.

It follows from this that $c = m$. Also, along S_3, we have that

$$\sum_{i=1}^{m} a_i = \lim \sum_{i=1}^{m} a_i^{n+1} = m.$$

Therefore $\sum_{i=1}^{m} \operatorname{Re} a_i = m$. But the real part of each a_i does not exceed 1, with equality if and only if $a_i = 1$, it follows that $a_i = 1$ for each i.

[Problem **2010:5** was contributed by Bamdad R. Yahaghi.]

2010:6. Let $f(x)$ be a quadratic polynomial. Prove that there exist quadratic polynomials $g(x)$ and $h(x)$ for which

$$f(x)f(x+1) = g(h(x)).$$

Solution 1. [A. Remorov] Let $f(x) = a(x - r)(x - s)$. Then

$$f(x)f(x+1) = a^2(x - r)(x - s + 1)(x - r + 1)(x - s)$$
$$= a^2(x^2 + x - rx - sx + rs - r)(x^2 + x - rx - sx + rs - s)$$
$$= a^2[(x^2 - (r + s - 1)x + rs) - r][(x^2 - (r + s - 1)x + rs) - s]$$
$$= g(h(x)),$$

where $g(x) = a^2(x - r)(x - s) = af(x)$ and $h(x) = x^2 - (r + s - 1)x + rs$.

Solution 2. Let $f(x) = ax^2 + bx + c$, $g(x) = px^2 + qx + r$ and $h(x) = ux^2 + vx + w$. Then

$$f(x)f(x+1) = a^2x^4 + 2a(a + b)x^3 + (a^2 + b^2 + 3ab + 2ac)x^2 + (b + 2c)(a + b)x$$
$$+ c(a + b + c)$$
$$g(h(x)) = p(ux^2 + vx + w)^2 + q(ux^2 + vx + w) + r$$
$$= pu^2x^4 + 2puvx^3 + (2puw + pv^2 + qu)x^2$$
$$+ (2pvw + qv)x + (pw^2 + qw + r).$$

Equating coefficients, we find that $pu^2 = a^2$, $puv = a(a+b)$, $2puw+pv^2+qu = a^2+b^2+3ab+2ac$, $(b+2c)(a+b) = (2pw+q)v$ and $c(a+b+c) = pw^2+qw+r$. We need to find just one solution of this system. Let $p = 1$ and $u = a$. Then $v = a + b$ and $b + 2c = 2pw + q$ from the second and fourth equations. This yields the third equation automatically. Let $q = b$ and $w = c$. Then from the fifth equation, we find that $r = ac$.

Thus, when $f(x) = ax^2 + bx + c$, we can take $g(x) = x^2 + bx + ac$ and $h(x) = ax^2 + (a + b)x + c$.

Solution 3. [S. Wang] Suppose that

$$f(x) = a(x + h)^2 + k = a(t - (1/2))^2 + k,$$

where $t = x + h + \frac{1}{2}$. Then $f(x + 1) = a(x + 1 + h)^2 + k = a(t + (1/2))^2 + k$, so that

$$f(x)f(x + 1) = a^2(t^2 - (1/4))^2 + 2ak(t^2 + (1/4)) + k^2$$

$$= a^2t^4 + \left(-\frac{a^2}{2} + 2ak\right) t^2 + \left(\frac{a^2}{16} + \frac{ak}{2} + k^2\right).$$

Thus, we can achieve the desired representation with $h(x) = t^2 = x^2 + (2h + 1)x + \frac{1}{4}$ and $g(x) = a^2x^2 + (\frac{-a^2}{2} + 2ak)x + (\frac{a^2}{16} + \frac{ak}{2} + k^2)$.

Solution 4. [V. Krakovna] Let $f(x) = ax^2 + bx + c = au(x)$ where $u(x) = x^2 + dx + e$, where $b = ad$ and $c = ae$. If we can find functions $v(x)$ and $w(x)$ for which $u(x)u(x+1) = v(w(x))$, then $f(x)f(x+1) = a^2v(w(x))$, and we can take $h(x) = w(x)$ and $g(x) = a^2v(x)$.

Define $p(t) = u(x + t)$, so that $p(t)$ is a monic quadratic in t. Then, noting that $p''(t) = u''(x + t) = 2$, we have that

$$p(t) = u(x + t) = u(x) + u'(x)t + \frac{u''(x)}{2}t^2 = t^2 + u'(x)t + u(x),$$

from which we find that

$$u(x)u(x + 1) = p(0)p(1) = u(x)[u(x) + u'(x) + 1]$$

$$= u(x)^2 + u'(x)u(x) + u(x) = p(u(x)) = u(x + u(x)).$$

Thus, $u(x)u(x + 1) = v(w(x))$ where $w(x) = x + u(x)$ and $v(x) = u(x)$. Therefore, we get the desired representation with

$$h(x) = x + u(x) = x^2 + \left(1 + \frac{b}{a}\right) x + \frac{c}{a}$$

and

$$g(x) = a^2v(x) = a^2u(x) = af(x) = a^2x^2 + abx + ac.$$

Comments. The second solution can also be obtained by looking at special cases, such as when $a = 1$ or $b = 0$, getting the answer and then making a conjecture.

There is an interesting story behind this problem. Originally it was noted that the product of two consecutive oblong numbers (products of consecutive integers) was also oblong, since for each integer n, $(n - 1)n \times n(n + 1) = (n^2 - 1)n^2$. The same is true for integers one more than a perfect square, since $[(n - 1)^2 + 1][n^2 + 1] = (n^2 - n + 1)^2 + 1$, and numbers of the form $n^2 + n + 1$ since $(n^2 - n + 1)(n^2 + n + 1) = n^4 + n^2 + 1$. It was not difficult to conjecture and prove the result that $f(x)f(x+1) = f(x+f(x))$ for monic quadratics $f(x)$.

This problem was also given to high school students participating in a problems correspondence program. One solver, James Rickards, then in Grade 9 at an Ottawa, Canada high school, noted that, if the roots of $f(x)$ were r and s, then those of $f(x+1)$ were $r-1$ and $s-1$. Thus, $f(x)f(x+1)$ was a quartic polynomial two of whose roots had the same sum as the other two. He proved that this root property characterized quartic polynomials representable as the composite of two quadratics.

In fact, he generalized to polynomials of higher degree and showed that a polynomial of degree mn could be written as the composite $g(h(x))$ of polynomials g of degree m and h of degree n if and only if its set of roots could be partitioned into m sets of n elements (not necessarily distinct) for which all the fundamental symmetric functions of degrees up to $m-1$ had the same values. I was unable to find any indication of this result in the literature, and so felt that it should be published in a journal with wide circulation. It was accepted and published under the title: James Rickards, *When is a polynomial a composition of other polynomials? American Mathematical Monthly* 118:4 (April, 2011), 358–363.

2011:5. Solve the system

$$x + xy + xyz = 12$$
$$y + yz + yzx = 21$$
$$z + zx + zxy = 30.$$

Solution 1. Let $u = xyz$. Then $x + xy = 12 - u$ so that $z + z(12 - u) = 30$ and $z = 30/(13 - u)$. Similarly, $y = 21/(31 - u)$ and $x = 12/(22 - u)$. Plugging these expressions (with $xyz = u$) into any one of the three equations yields that

$$0 = u^3 - 65u^2 + 1306u - 7560 = (u - 10)(u - 27)(u - 28) .$$

We get the three solutions

$$(x, y, z) = (1, 1, 10), \left(-\frac{12}{5}, \frac{21}{4}, -\frac{15}{7}\right), (-2, 7, -2),$$

all of which satisfy the system.

Solution 2. Define u and obtain the expressions for x, y, z as in Solution 1. Substitute these into the equation $xyz = u$. This leads to the quartic equation

$$
\begin{aligned}
0 &= u(u-13)(u-22)(u-31) + (12)(21)(30) \\
&= u^4 - 66u^3 + 1371u^2 - 8866u + 7560 \\
&= (u-1)(u^3 - 65u^2 + 1306u - 7560).
\end{aligned}
$$

Apart from the three values of u already identified, we have $u = 1$. This leads to

$$
(x, y, z) = \left(\frac{4}{7}, \frac{7}{10}, \frac{5}{2} \right).
$$

While, indeed, $xyz = 1$, we find that $x + xy + xyz = 69/35$, $y + yz + xyz = 69/20$, $z + zx + xyz = 69/14$, so the solution $u = 1$ is extraneous.

[Problem **2011:5** was posed by Stanley Rabinowitz in the Spring, 1982 issue of *AMATYC Review.*]

2011:6. [The problem posed on the competition is not available for publication, as it appeared as Enigma problem 1610 in the August 25, 2010 issue of the *New Scientist.* It concerned the determination of the final score of a three game badminton match where the score of each competitor was in arithmetic progression. The full statement of the problem is available on the internet.]

Solution 1. [K. Ng] Recall the winner of a badminton game must obtain at least 21 points and lead the opponent by at least 2 points. Let the scores for the three games be $a, a+u, a+2u$ for player A and $b, b+v, b+2v$ for player B. Suppose, wolog, that A wins the first game and that B wins the second. Then $a - b \geq 2$ and $(b+v) - (a+u) \geq 2$, whence $v - u \geq 2 + (a-b) \geq 4$. Hence

$$
(b+2v) - (a+2u) = [(b+v) - (a+u)] + (v-u) \geq 6 .
$$

Thus the third game is won by B by at least 6 points. Therefore, B must have exactly 21 points in the third game. It follows that B has more than 21 points in each of the previous games, as A does in the first game.

Hence $a - b = 2$ and

$$
2 \neq (b+v) - (a+u) = (v-u) - (a-b) = (v-u) - 2,
$$

so that $v - u = 4$. Therefore, the difference in the scores of the third game is exactly $2(v-u) - (a-b) = 6$ and the final scores are $(15, 21)$.

Solution 2. There are two different ways the score can present at the conclusion of a game: (a) The winner can have 21 points and the other no more than 19 points; (b) The winner can have at least 22 points and the loser two points fewer than the winner.

It is not possible for the scores in the first and third games to be of type (b). For the signed differences of the scores is also an arithmetic progression of three numbers, the first and third of which are both 2, both -2, or 2 and

−2. In the first two cases, this would mean that the same person won all three games, and in the third case that the scores in the second game were the same and there would be no winner of the second game.

Therefore either the opening or closing game has one person with a score of 21 and the other with a score s less than 20. Let the scores of the first player be $(21, 21+x, 21+2x)$ and of the second be $(s, s+y, s+2y)$, possibly in reverse order. Note that x cannot be negative, since otherwise $s \leq s + 2y = 21 \leq s + y$. Since not all three games are won by the same person, we must have $y > x$, $s + y + 2 = 21 + x$ and $21 + 2x + 2 = s + 2y$. Therefore $s = 19 - (y - x) = 23 - 2(y - x)$, whence $s = 15$. Since the winner of the game with scores $(21, 15)$ is the same as that of the second game and different from that of the third, this game must be the final game of the match. Therefore, the scores of the final game are $(21, 15)$.

Note that $y - x = 4$, and that, whenever $x \geq 1$, sets of scores $(21 + 2x, 23 + 2x), (21 + x, 19 + x), (21, 15)$ all satisfy the conditions.

2011:10. Suppose that p is an odd prime. Determine the number of subsets S contained in $\{1, 2, \ldots, 2p - 1, 2p\}$ for which (a) S has exactly p elements, and (b) the sum of the elements of S is a multiple of p.

Solution. Let ζ be a primitive pth root of unity, so that ζ is a power of $\cos(2\pi/p) + i\sin(2\pi/p)$ with the exponent not divisible by p. Let

$$f(x) = \prod_{i=1}^{2p}(x - \zeta^i) = (x^p - 1)^2 = x^{2p} - 2x^p + 1.$$

We can also write

$$f(x) = x^{2p} + \cdots \left[\sum \{(-1)^p \zeta^{i_1 + \cdots + i_p} : \{i_1, i_2, \cdots, i_p\} \right.$$

$$\left. \subseteq \{1, 2, \ldots, 2p\} \right] x^p + \cdots + 1$$

$$= x^{2p} + \cdots - (a_0 + a_1\zeta + a_2\zeta^2 + \cdots + a^{p-1}\zeta^{p-1})x^p + \cdots + 1,$$

for some coefficients a_i. We are required to determine the value of a_0.

Let $g(x) = (a_0 - 2) + a_1 x + a_2 x^2 + \cdots + a_{p-1}x^{p-1}$. Observe that $g(x)$ vanishes when x is any primitive pth root of unity and that the sum defining the coefficient of x^p has $\binom{2p}{p}$ terms, so that $a_0 + a_1 + \cdots + a_{p-1} = \binom{2p}{p}$. Since $g(x)$ is a polynomial of degree $p - 1$ whose zeros are the primitive pth roots of unity, $g(x)$ must be a constant multiple of $1 + x + \cdots + x^{p-1}$, so that

$$a_0 - 2 = a_1 = a_2 = \cdots = a_{p-1}.$$

Thus $a_0 + (p - 1)(a_0 - 2) = \binom{2p}{p}$, so that

$$a_0 = \frac{1}{p}\left[\binom{2p}{p} - 2 \right] + 2.$$

[Problem **2011:10** was contributed by Ali Feiz Mohammadi.]

2012:2. Suppose that f is a function defined on the set $\mathbb{Z}$ of integers that takes integer values and satisfies the condition that $f(b) - f(a)$ is a multiple of $b - a$ for every pair a, b, of integers. Suppose also that p is a polynomial with integer coefficients such that $p(n) = f(n)$ for infinitely many integers n. Prove that $p(x) = f(x)$ for every positive integer x.

The solution to this problem appears in Chap. 11.

2012:3. Given the real numbers a, b, c not all zero, determine the real solutions x, y, z, u, v, w for the system of equations:

$$x^2 + v^2 + w^2 = a^2$$
$$u^2 + y^2 + w^2 = b^2$$
$$u^2 + v^2 + z^2 = c^2$$
$$u(y + z) + vw = bc$$
$$v(x + z) + wu = ca$$
$$w(x + y) + uv = ab.$$

Solution 1. Evaluating $b^2 c^2$ in two ways, we find that
$$0 = (u^2 + y^2 + w^2)(u^2 + v^2 + z^2) - (uy + uz + vw)^2$$
$$= (u^2 - yz)^2 + (uv - wz)^2 + (wu - vy)^2.$$

Hence $u^2 = yz$. Similarly, $v^2 = zx$ and $w^2 = xy$. Inserting these three values into the first three equations yields that
$$x(x + y + z) = a^2$$
$$y(x + y + z) = b^2$$
$$z(x + y + z) = c^2.$$

It follows that not all of x, y, z can vanish and that $(x+y+z)^2 = a^2+b^2+c^2$. Also
$$x : a^2 = y : b^2 = z : c^2 = 1 : (x+y+z) = (x+y+z) : (a^2 + b^2 + c^2).$$

Therefore
$$(x, y, z, u, v, w) = (a^2 d, b^2 d, c^2 d, bcd, cad, abd)$$
where $d^2(a^2 + b^2 + c^2) = 1$.

Solution 2. [Y. Wu; P.J. Zhao] From the given equations, we see that
$$(x, v, w) \cdot (w, u, y) = w(x + y) + uv = ab = \sqrt{x^2 + v^2 + w^2}\sqrt{w^2 + u^2 + y^2};$$
$$(w, u, y) \cdot (v, z, u) = u(y + z) + vw = bc = \sqrt{w^2 + u^2 + y^2}\sqrt{v^2 + z^2 + u^2};$$
$$(v, z, u) \cdot (x, v, w) = v(x + z) + uw = ca = \sqrt{v^2 + z^2 + u^2}\sqrt{x^2 + v^2 + w^2}.$$

It follows from the conditions for equality in the Cauchy-Schwarz Inequality that (x, v, w), (w, u, y) and (v, z, u) are all constant multiples of the same unit vector (p, q, r). Indeed

$$(x, v, w) = a(p, q, r); \quad (w, u, y) = b(p, q, r); \quad (v, z, u) = c(p, q, r).$$

Thus, $x = ap$, $y = br$, $z = cq$, $u = bq = cr$, $v = aq = cp$ and $w = ar = bp$, whence

$$u^2 = bcqr = yz; \quad v^2 = acpq = xz; \quad w^2 = abpr = xy.$$

The solution now can be completed as in Solution 1.

2013:6. Let $p(x) = x^4 + ax^3 + bx^2 + cx + d$ be a polynomial with rational coefficients. Suppose that $p(x)$ has exactly one real root r. Prove that r is rational.

Solution. Since nonreal roots occur in pairs, $p(x)$ must have an even number of real roots counting multiplicity. Therefore r must be a double or quadruple root. If r is a quadruple root, then $p(x) = (x - r)^4$ and $r = -a/4$ is rational. Suppose, otherwise, that $p(x) = (x - r)^2 q(x)$ where $q(x)$ is an irreducible quadratic. The derivative $p'(x)$ is equal to $(x - r)f(x)$ where $f(x) = 2q(x) + (x - r)q'(x)$. Since $q(r)$ does not vanish, $f(r) \neq q(r)$ so that $f(x)$ and $q(x)$ must be distinct and coprime. The monic greatest common divisor of $p(x)$ and $p'(x)$ must therefore be $x - r$. Since (by the Euclidean algorithm) this is a polynomial with rational coefficients, therefore r is rational.

2014:6. Let $f(x) = x^6 - x^4 + 2x^3 - x^2 + 1$.

(a) Prove that $f(x)$ has no positive real roots.

(b) Determine a nonzero polynomial $g(x)$ of minimum degree for which all the coefficients of $f(x)g(x)$ are nonnegative rational numbers.

(c) Determine a polynomial $h(x)$ of minimum degree for which all the coefficients of $f(x)h(x)$ are positive rational numbers.

Solution. (a) Note that

$$f(x) = (x^2 - 1)(x^4 - 1) + 2x^3 = (x^2 - 1)^2(x^2 + 1) + 2x^3$$

from which we see that $f(x) > 0$ for all $x > 0$. Alternatively,

$$f(x) = x^6 + x^3 + (x^3 - x^4) + (1 - x^2) = x^4(x^2 - 1) + x^2(x - 1) + x^3 + 1$$

from which we see that there are no roots in either of the intervals $[0, 1]$ and $[1, \infty)$.

(b) It is straightforward to see that multiplying $f(x)$ by a linear polynomial will not achieve the goal. We have that

$$f(x)(x^2 + bx + c) = x^8 + bx^7 + (c - 1)x^6 + (2 - b)x^5 + (2b - c - 1)x^4$$
$$+ (2c - b)x^3 + (1 - c)x^2 + bx + c$$

from which we see that taking $c = 1$ and $1 \leq b \leq 2$ will yield the desired polynomial $g(x)$. Examples of suitable $g(x)$ are $x^2 + x + 1$ and $(x + 1)^2$.

(c) From (b), we note that no quadratic polynomial will serve. However, taking $h(x) = x^3 + 2x^2 + 2x + 1$, we find that

$$f(x)h(x) = x^9 + 2x^8 + x^7 + x^6 + x^5 + x^4 + x^3 + x^2 + 2x + 1.$$

Comment. It is known that for a given real polynomial $f(x)$, there exists a polynomial $g(x)$ for which all the coefficients of $f(x)g(x)$ are nonnegative (resp. positive) if and only if none of the roots of $f(x)$ are positive (resp. nonnegative).

Another example is $f(x) = x^3 - x + 1$. Since $f(x) = x^3 + (1 - x) = x(x^2 - 1) + 1$, we see that there are no roots in $[0, 1]$ and $[1, \infty)$. No linear polynomial $x + c$ will do for $g(x)$. Since

$$f(x)(x^2 + bx + c) = x^5 + bx^4 + (c - 1)x^3 + (1 - b)x^2 + (b - c)x + c,$$

we require that $1 \le c \le b$ and $0 \le b \le 1$. The only possibility is that $g(x) = x^2 + x + 1$.

For strictly positive coefficients in the product, set $g(x) = ax^3 + bx^2 + cx + d$. Then

$$f(x)g(x) = ax^6 + bx^5 + (c - a)x^4 + (a + d - b)x^3 + (b - c)x^2 + (c - d)x + d.$$

For g to be suitable, we require that $c > a > 0$, $a + d > b$ and $b > c > d > 0$. We can take $g(x) = 3x^3 + 5x^2 + 4x + 3$.

[The examples in Problem **2014:6** are due to Horst Brunotte in Düsseldorf, Germany.]

2015:10. (a) Let

$$g(x, y) = x^2 y + xy^2 + xy + x + y + 1.$$

We form a sequence $\{x_0\}$ as follows: $x_0 = 0$. The next term x_1 is the unique solution -1 of the linear equation $g(t, 0) = 0$. For each $n \ge 2$, x_n is the solution other than x_{n-2} of the equation $g(t, x_{n-1}) = 0$.

Let $\{f_n\}$ be the Fibonacci sequence determined by $f_0 = 0$, $f_1 = 1$ and $f_n = f_{n-1} + f_{n-2}$ for $n \ge 2$. Prove that, for any nonnegative integer k,

$$x_{2k} = \frac{f_k}{f_{k+1}} \qquad \text{and} \qquad x_{2k+1} = -\frac{f_{k+2}}{f_{k+1}}.$$

(b) Let

$$h(x, y) = x^2 y + xy^2 + \beta xy + \gamma(x + y) + \delta$$

be a polynomial with real coefficients β, γ, δ. We form a bilateral sequence $\{x_n : n \in \mathbf{Z}\}$ as follows. Let $x_0 \ne 0$ be given arbitrarily. We select x_{-1} and x_1 to be the two solutions of the quadratic equation $h(t, x_0) = 0$ in either order. From here, we can define inductively the terms of the sequence for positive and negative values of the index so that x_{n-1} and x_{n+1} are the two solutions of the equation $h(t, x_n) = 0$. We suppose that at each stage, neither of these solutions is zero.

Prove that the sequence $\{x_n\}$ has period 5 (i.e. $x_{n+5} = x_n$ for each index n) if and only if $\gamma^3 + \delta^2 - \beta\gamma\delta = 0$.

(a) *Solution*. Observe that

$$g(x, y) = (xy + 1)(x + y + 1).$$

For each value of $y \neq 0$, the equation $g(x, y) = 0$ has two solutions: $y = -1/x$ and $y = -(x + 1)$. Observe that $g(x, y)$ is symmetrical in x and y, so that, for each consecutive pair x_n, x_{n+1} of terms in the sequence $g(x_n, x_{n+1}) = g(x_{n+1}, x_n) = 0$. Consider the equation $0 = g(t, x_1) = g(t, -1) = (-t + 1)t$. One of its solutions is $x_0 = 0$ and the other is $x_2 = 1$.

For the equation $0 = g(t, x_2)$, we have that $g(x_1, x_2) = g(x_2, x_1) = 0$ and $x_1 x_2 + 1 = 0$. Therefore $x_2 + x_3 + 1 = 0$, so that $x_3 = -(x_2 + 1)$. Continuing on in this way, we find that, for each positive integer k, $x_{2k-1} x_{2k} = -1$ and $x_{2k} + x_{2k+1} = -1$, whereupon

$$x_{2k+1} = -1 + \frac{1}{x_{2k-1}} = \frac{1 - x_{2k-1}}{x_{2k-1}}.$$

When $k = 1$, we find that $x_{2k-1} = x_1 = -1 = -f_2/f_1$. Suppose, for $k \geq 1$, we have that $x_{2k-1} = -f_{k+1}/f_k$. Then

$$x_{2k+1} = -1 - \frac{f_k}{f_{k+1}} = -\frac{f_{k+1} + f_k}{f_{k+1}} = -\frac{f_{k+2}}{f_{k+1}}.$$

By induction, we obtain the desired expression for x_{2k+1}. Also $x_{2k} = -1/x_{2k-1} = f_k/f_{k+1}$.

(b) *Solution 1*. Observe that from the sum of the roots of $h(t, x_n) = 0$, we have that

$$x_{n-1} + x_{n+1} = -\left[\frac{x_n^2 + \beta x_n + \gamma}{x_n}\right],$$

or

$$x_{n-1} + x_n + x_{n+1} = -\beta - \frac{\gamma}{x_n}$$

for each n. The sequence will have period 5 if and only if the sum of any five consecutive terms is constant.

Since, for each integer n,

$$x_{n+2} + x_{n+1} + x_n = -\beta - \frac{\gamma}{x_{n+1}}$$

and

$$x_n + x_{n-1} + x_{n-2} = -\beta - \frac{\gamma}{x_{n-1}},$$

we have that

$$x_{n+2} + x_{n+1} + x_n + x_{n-1} + x_{n-2} = -2\beta - \gamma\left(\frac{x_{n+1} + x_{n-1}}{x_{n+1} x_{n-1}}\right) - x_n$$

$$= -2\beta + \gamma\left(\frac{x_n^2 + \beta x_n + \gamma}{\gamma x_n + \delta}\right) - x_n$$

$$= \frac{-(\beta\gamma + \delta)x_n + (\gamma^2 - 2\beta\delta)}{\gamma x_n + \delta}$$

$$= -\left(\beta + \frac{\delta}{\gamma}\right) + \left(\frac{\gamma^3 + \delta^2 - \beta\gamma\delta}{\gamma^2 x_n + \gamma\delta}\right).$$

This sum is independent of n if and only if the term involving x_n vanishes identically, i.e., if and only if the required condition holds.

Solution 2. From the formula for the product of the roots, we obtain that

$$x_{n-1}x_{n+1} = \frac{\gamma x_n + \delta}{x_n}$$

so that

$$x_{n-1}x_n x_{n+1} = \gamma x_n + \delta$$

for each index n. Therefore

$$x_{n+2}x_{n+1}x_n^2 x_{n-1}x_{n-2} = (\gamma x_{n+1} + \delta)(\gamma x_{n-1} + \delta)$$

$$= \gamma^2\left(\frac{\gamma x_n + \delta}{x_n}\right) - \gamma\delta\left(x_n + \beta + \frac{\gamma}{x_n}\right) + \delta^2$$

$$= (\gamma^3 + \delta^2 - \beta\gamma\delta) - \gamma\delta x_n$$

whence

$$x_{n+2}x_{n+1}x_n x_{n-1}x_{n-2} = -\gamma\delta + \left(\frac{\gamma^3 + \delta^2 - \beta\gamma\delta}{x_n}\right).$$

The result again follows.

Comments. If $x_0 = 0$, then $h(t, x_0)$ is linear and there is a single root $-\delta/\gamma$. We can extend the sequence in only one direction, and it begins with the terms $0, -\delta/\gamma, (\gamma^3 + \delta^2 - \beta\gamma\delta)/(\gamma\delta), \ldots$.

The study of the type of sequence given in this problem began with the recursions given by

$$x_{n+1} = \frac{x_n + c}{x_{n-1}}$$

or, equivalently, if you want to continue the sequence backwards,

$$x_{n-1} = \frac{x_n + c}{x_{n+1}},$$

where c is a parameter and n ranges over the integers. We suppose that the terms of the sequence are selected to avoid division by 0. When $c = 0$ and $c = 1$, any bilateral sequence satisfying the recursion is periodic with respective periods 6 and 5. For other values of c, certain but not all sequences are periodic. The examination of the periodic cases lead to the identification of an invariant for the sequence.

When

$$f_c(x, y) = \frac{x^2 y + xy^2 + x^2 + (c+1)x + y^2 + (c+1)y + c}{xy},$$

it turns out that

$$f(x, y) = f\left(y, \frac{y+c}{x}\right)$$

with the result that $f(x_n, x_{n+1})$ is constant with respect to the index n.

The equation $f_c(x, y) = k$ can be rewritten as $h_{c,k}(x, y) = 0$ where

$$\begin{aligned} h_{c,k}(x, y) &= x^2 y + xy^2 + x^2 + y^2 - kxy + (c+1)(x+y) + c \\ &= xy(x+y) + (x+y)^2 - (k+2)xy + (c+1)(x+y) + c \\ &= (y+1)x^2 + (y^2 - ky + (c+1))x + (y+1)(y+c) \ . \end{aligned}$$

The function $h_{c,k}(x, y)$ is a symmetric polynomial quadratic in each variable. For each n, the terms x_{n-1} and x_{n+1} are the roots of the quadratic equation $h_{c,k}(x, x_n) = 0$, provided $x_n \neq 1$. Indeed, from the relation between the coefficients and the product of the roots of a quadratic, we corroborate the relation

(2.13) $$x_{n-1}x_{n+1} = x_n + c.$$

Moreover, we have

(2.14) $$x_{n-1} + x_{n+1} = \frac{x_n^2 - kx_n + c + 1}{x_n + 1}.$$

These sequences, for various values of the parameter c, are studied in the paper Ed Barbeau, Boaz Gelbord and Steve Tanny, *Periodicities of solutions of the generalized Lyness recursion*, J. *Difference Equations and Applications* 1 (1995), 291–306.

We look at a slight generalization and point out how two types of recursions are related to each other. Let

$$\begin{aligned} h(x, y) &= x^2 y + xy^2 + \alpha(x^2 + y^2) + \beta xy + \gamma(x+y) + \delta \\ &= (y+\alpha)x^2 + (y^2 + \beta y + \gamma)x + (\alpha y^2 + \gamma y + \delta). \end{aligned}$$

As before, we can seed the recursion by specifying x_0 and arranging that, for each integer n, x_{n-1} and x_{n+1} are the solutions of the equation $h(x, x_n) = 0$. To avoid complications, we suppose that x_n never assumes the value $-\alpha$.

Sequences determined in this way satisfy both the recursions:

(2.15) $$x_{n+1} + x_{n-1} = -\left[\frac{x_n^2 + \beta x_n + \gamma}{x_n + \alpha}\right]$$

and

(2.16) $$x_{n+1}x_{n-1} = \frac{\alpha x_n^2 + \gamma x_n + \delta}{x_n + \alpha}.$$

Remarkably, if we are given any sequence that satisfies the first of these recursion relations, we can select a constant δ so that it satisfies the second. Similarly, given any sequences given by the second recursion relationship, we can select a value of β so that it satisfies the first. Thus, either recursion individually allows us to introduce a function $h(x,y)$ that will allow us to generate the sequence. This follows from the following proposition.

We have

$$h\left(\frac{\alpha y^2 + \gamma y + \delta}{x(y+\alpha)}, y\right) = \frac{\alpha y^2 + \gamma y + \delta}{x^2(y+\alpha)}h(x,y)$$

and

$$h\left(-\left[\frac{\alpha y^2 + \beta y + \gamma}{y+\alpha}\right] - x, y\right) = h(x,y)$$

with the result that, when x and y are consecutive terms, $\frac{h(x,y)}{xy}$ is invariant along any recursion satisfying (2.16) and $h(x,y)$ is invariant along any recursion satisfying (2.15). Thus, any recursion satisfying either (2.15) or (2.16) will satisfy the other with a suitable choice of parameters.

Proof.

$$h\left(\frac{\alpha y^2 + \gamma y + \delta}{x(y+\alpha)}, y\right) = (y+\alpha)\frac{(\alpha y^2 + \gamma y + \delta)^2}{x^2(y+\alpha)^2} + (y^2 + \beta y + \gamma)\frac{\alpha y^2 + \gamma y + \delta}{x(y+\alpha)}$$
$$+ (\alpha y^2 + \gamma y + \delta)$$

$$= \frac{\alpha y^2 + \gamma y + \delta}{x^2(y+\alpha)}\left[(\alpha y^2 + \gamma y + \delta) + x(y^2 + \beta y + \gamma) + x^2(y+\alpha)\right]$$

$$= \frac{\alpha y^2 + \gamma y + \delta}{x^2(y+\alpha)}\left[x^2 y + y^2 x + \alpha(x^2 + y^2) + \beta xy + \gamma(x+y) + \delta\right]$$

$$= \frac{\alpha y^2 + \gamma y + \delta}{x^2(y+\alpha)}h(x,y).$$

Also

$$h\left(-\left[\frac{y^2 + \beta y + \gamma}{y+\alpha}\right] - x, y\right)$$

$$= (y+\alpha)\left[x^2 + \frac{2x(y^2 + \beta y + \gamma)}{y+\alpha} + \frac{(y^2 + \beta y + \gamma)^2}{(y+\alpha)^2}\right]$$

$$- (y^2 + \beta y + \gamma)\left[\frac{y^2 + \beta y + \gamma}{y+\alpha} + x\right] + (\alpha y^2 + \gamma y + \delta)$$

$$= x^2 y + \alpha x^2 + 2xy^2 + 2\beta xy + 2\gamma x - xy^2 - \beta xy - \gamma x + \alpha y^2 + \gamma y + \delta$$

$$= h(x,y).$$

CHAPTER 3

Inequalities

2001:1. Let $a, b, c > 0$, $a < bc$ and $1 + a^3 = b^3 + c^3$. Prove that $1 + a < b + c$.

Solution 1. Since $(1 + a)(1 - a + a^2) = (b + c)(b^2 - bc + c^2)$, and since $1 - a + a^2$ and $b^2 - bc + c^2 = \frac{1}{2}(b - c)^2 + \frac{1}{2}(b^2 + c^2)$ are positive, we have that

$$1 + a < b + c \Leftrightarrow 1 - a + a^2 > b^2 - bc + c^2.$$

Suppose, if possible, that $1 + a \geq b + c$. Then

$$b^2 - bc + c^2 \geq 1 - a + a^2$$
$$\Rightarrow (b + c)^2 - 3bc \geq (1 + a)^2 - 3a > (1 + a)^2 - 3bc$$
$$\Rightarrow (b + c)^2 > (1 + a)^2 \Rightarrow b + c > 1 + a$$

which is a contradiction.

Solution 2. [J. Chui] Let $u = (1 + a) - (b + c)$. Then

$$(1 + a)^3 - (b + c)^3 = u[(1 + a)^2 + (1 + a)(b + c) + (b + c)^2]$$
$$= u[(1 + a)^2 + (1 + a)(b + c) + b^2 + 2bc + c^2].$$

But also

$$(1 + a)^3 - (b + c)^3 = (1 + a^3) - (b^3 + c^3) + 3a(1 + a) - 3bc(b + c)$$
$$= 0 + 3[a(1 + a) - bc(b + c)] < 3bcu.$$

It follows from these that

$$0 > u[(1 + a)^2 + (1 + a)(b + c) + b^2 - bc + c^2].$$

Since the quantity in square brackets is positive, we must have that $u < 0$, as desired.

© Springer International Publishing Switzerland 2016

E.J. Barbeau, *University of Toronto Mathematics Competition (2001–2015)*, Problem Books in Mathematics,

DOI 10.1007/978-3-319-28106-3_3

Solution 3. [A. Momin; N. Martin] Suppose, if possible, that $(1 + a) \geq (b + c)$. Then

$$0 \leq (1+a)^2 - (b+c)^2 = (1+a^2) - (b^2+c^2) - 2(bc-a) < (1+a^2) - (b^2+c^2).$$

Hence $1 + a^2 > b^2 + c^2$. It follows that

$$(1 - a + a^2) - (b^2 - bc + b^2) = (1 + a^2) - (b^2 + c^2) + (bc - a) > 0$$

so that

$$(1 - a + a^2) > (b^2 - bc + c^2).$$

However

$$(1 + a)(1 - a + a^2) = 1 + a^3 = b^3 + c^3 = (b + c)(b^2 - bc + c^2),$$

from which it follows that $1 + a < b + c$, yielding a contradiction. Hence, the desired result follows.

Solution 4. [P. Gyrya] Let $p(x) = x^3 - 3ax$. Checking the first derivative yields that $p(x)$ is strictly increasing for $x > \sqrt{a}$. Now $1 + a \geq 2\sqrt{a} > \sqrt{a}$ and $b + c \geq 2\sqrt{bc} > 2\sqrt{a} > \sqrt{a}$, so both $1 + a$ and $b + c$ lie in the part of the domain of $p(x)$ where it strictly increases. Now

$$p(1 + a) = (1 + a)^3 - 3a(1 + a) = 1 + a^3 = b^3 + c^3$$
$$= (b + c)^3 - 3bc(b + c) < (b + c)^3 - 3a(b + c) = p(b + c)$$

from which it follows that $1 + a < b + c$.

Solution 5. Consider the function $g(x) = x(1 + a^3 - x) = x(b^3 + c^3 - x)$. Then $g(1) = g(a^3) = a^3$ and $g(b^3) = g(c^3) = (bc)^3$. Since $a^3 < (bc)^3$ and the graph of $g(x)$ is a parabola opening down, it follows that b^3 and c^3 lie between 1 and a^3.

Now consider the function $h(x) = x^{1/3} + (b^3 + c^3 - x)^{1/3} = x^{1/3} + (1 + a^3 - x)^{1/3}$ for $0 \leq x \leq 1 + a^3$. Then $h(1) = h(a^3) = 1 + a$ and $h(b^3) = h(c^3) = b + c$. Since

$$h''(x) = -\frac{2}{9}(x^{-5/3} + (1 + a^3 - x)^{-5/3}),$$

the function $h(x)$ is concave. Since b^3 and c^3 lie between 1 and a^3, it follows that $1 + a < b + c$, as desired.

2002:6. Let $x, y > 0$ be such that $x^3 + y^3 \leq x - y$. Prove that $x^2 + y^2 \leq 1$.

Solution 1. Let $y = tx$. Since $x > y > 0$, we have that $0 < t < 1$. Then $x^3(1 + t^3) \leq x(1 - t) \Rightarrow x^2(1 + t^3) \leq (1 - t)$, Therefore,

$$x^2 + y^2 = x^2(1 + t^2) \leq \left(\frac{1 - t}{1 + t^3}\right)(1 + t^2)$$

$$= \frac{1 - t + t^2 - t^3}{1 + t^3} = 1 - \frac{t(1 - t + 2t^2)}{1 + t^3}.$$

Since $1 - t + 2t^2$, having negative discriminant, is always positive, the desired result follows.

Solution 2. [J. Chui] Suppose, if possible, that $x^2 + y^2 = r^2 > 1$. We can write $x = r \sin \theta$ and $y = r \cos \theta$ for $0 < \theta < \pi/2$. Then

$$\begin{aligned}
x^3 + y^3 - (x - y) &= r^3 \sin^3 \theta + r^3 \cos^3 \theta - r \sin \theta + r \cos \theta \\
&> r \sin \theta (\sin^2 \theta - 1) + r \cos^3 \theta + r \cos \theta \\
&= -r \sin \theta \cos^2 \theta + r \cos^3 \theta + r \cos \theta \\
&= r \cos^2 \theta \left(\cos \theta + \frac{1}{\cos \theta} - \sin \theta \right) \\
&> r \cos^2 \theta (2 - \sin \theta) > 0,
\end{aligned}$$

contrary to hypothesis. The result follows by contradiction.

2003:2. Let a, b, c be positive real numbers for which $a + b + c = abc$. Prove that

$$\frac{1}{\sqrt{1 + a^2}} + \frac{1}{\sqrt{1 + b^2}} + \frac{1}{\sqrt{1 + c^2}} \le \frac{3}{2}.$$

Solution 1. Let $a = \tan \alpha$, $b = \tan \beta$, $c = \tan \gamma$, where $\alpha, \beta, \gamma \in (0, \pi/2)$. Then

$$\begin{aligned}
\tan(\alpha + \beta + \gamma) &= \frac{\tan \alpha + \tan \beta + \tan \gamma - \tan \alpha \tan \beta \tan \gamma}{1 - \tan \alpha \tan \beta - \tan \beta \tan \gamma - \tan \gamma \tan \alpha} \\
&= \frac{a + b + c - abc}{1 - ab - bc - ca} = 0,
\end{aligned}$$

whence $\alpha + \beta + \gamma = \pi$. Then, the left side of the inequality is equal to

$$\begin{aligned}
\cos \alpha + \cos \beta + \cos \gamma &= \cos \alpha + \cos \beta - \cos(\alpha + \beta) \\
&= 2 \cos \left(\frac{\alpha + \beta}{2} \right) \cos \left(\frac{\alpha - \beta}{2} \right) - 2 \cos^2 \left(\frac{\alpha + \beta}{2} \right) + 1 \\
&\le 2 \cos \left(\frac{\alpha + \beta}{2} \right) - 2 \cos^2 \left(\frac{\alpha + \beta}{2} \right) + 1 \\
&= \frac{3}{2} - \frac{1}{2} \left(2 \cos \left(\frac{\alpha + \beta}{2} \right) - 1 \right)^2 \le \frac{3}{2},
\end{aligned}$$

with equality if and only if $\alpha = \beta = \gamma = \pi/3$.

Solution 2. Define α, β and γ and note that $\alpha + \beta + \gamma = \pi$ as in Solution 1. Since $\cos x$ is a concave function on $[0, \pi/2]$, we have from Jensen's inequality that

$$\frac{\cos \alpha + \cos \beta + \cos \gamma}{3} \le \cos \left(\frac{\alpha + \beta + \gamma}{3} \right) = \cos \frac{\pi}{3} = \frac{1}{2},$$

from which the result follows.

2004:1. Prove that, for any nonzero complex numbers z and w,

$$(|z| + |w|) \left| \frac{z}{|z|} + \frac{w}{|w|} \right| \le 2|z + w|.$$

The solution to this problem appears in Chap. 2.

2004:4. Let n be a positive integer exceeding 1. How many permutations $\{a_1, a_2, \ldots, a_n\}$ of $\{1, 2, \ldots, n\}$ are there which maximize the value of the sum

$$|a_2 - a_1| + |a_3 - a_2| + \cdots + |a_{i+1} - a_i| + \cdots + |a_n - a_{n-1}|$$

over all permutations? What is the value of this maximum sum?

Solution. First, suppose that n is odd. Then

$$|a_{i+1} - a_i| \leq \left|a_{i+1} - \frac{n+1}{2}\right| + \left|a_i - \frac{n+1}{2}\right|$$

with equality if and only if $\frac{1}{2}(n+1)$ lies between a_{i+1} and a_i.

Hence

$$\sum_{i=1}^{n-1} |a_{i+1} - a_i| \leq 2 \left(\sum_{i=1}^{n} \left|a_i - \frac{n+1}{2}\right|\right) - \left(\left|a_1 - \frac{n+1}{2}\right| + \left|a_n - \frac{n+1}{2}\right|\right)$$

$$= \left(\sum_{i=1}^{n} |2a_i - (n+1)|\right) - \left(\left|a_1 - \frac{n+1}{2}\right| + \left|a_n - \frac{n+1}{2}\right|\right)$$

$$= [(n-1) + (n-3) + \cdots + 2 + 0 + 2 + \cdots + (n-1)]$$

$$- \left(\left|a_1 - \frac{n+1}{2}\right| + \left|a_n - \frac{n+1}{2}\right|\right)$$

$$= 4\left(1 + 2 + \cdots + \frac{n-1}{2}\right) - \left(\left|a_1 - \frac{n+1}{2}\right| + \left|a_n - \frac{n+1}{2}\right|\right)$$

$$= 4\left[\frac{((n-1)/2) \cdot ((n+1)/2)}{2}\right] - \left(\left|a_1 - \frac{n+1}{2}\right| + \left|a_n - \frac{n+1}{2}\right|\right)$$

$$\leq \frac{n^2 - 1}{2} - 1 = \frac{n^2 - 3}{2}$$

since $|a_1 - ((n+1)/2)| + |a_n - ((n+1)/2)| \geq 1$. Equality occurs when one of a_1 and a_n is equal to $\frac{1}{2}(n+1)$ and the other is equal to $\frac{1}{2}(n+1) \pm 1$.

We get a permutation giving the maximum value of $\frac{1}{2}(n^2 - 3)$ if and only if the foregoing conditions on a_1 and a_n are satisfied (in four possible ways) and $\frac{1}{2}(n+1)$ lies between a_i and a_{i+1} for each i. For example, if $a_1 = \frac{1}{2}(n+1) + 1$, then we require that the $\frac{1}{2}(n-3)$ numbers $a_3, a_5, \ldots, a_{n-2}$ exceed $\frac{1}{2}(n+1) + 1$ and the $\frac{1}{2}(n-1)$ numbers $a_2, a_4, \ldots, a_{n-1}$ are less than $\frac{1}{2}(n+1)$. Thus, there are

$$4\left(\frac{n-3}{2}\right)! \left(\frac{n-1}{2}\right)!$$

ways of achieving the maximum.

Now suppose that n is even. As before,

$$|a_{i+1} - a_i| \leq \left|a_{i+1} - \frac{n+1}{2}\right| + \left|a_i - \frac{n+1}{2}\right|$$

with equality if and only if $\frac{1}{2}(n+1)$ lies between a_{i+1} and a_i.

We have that

$$\sum_{i=1}^{n-1} |a_{i+1} - a_i| \leq 2(1 + 3 + \cdots + (n-1)) - \left[\left| a_1 - \frac{n+1}{2} \right| + \left| a_n - \frac{n+1}{2} \right| \right]$$

$$\leq \frac{n^2}{2} - 1 = \frac{n^2 - 2}{2},$$

with the latter inequality becoming equality if and only if $\{a_1, a_n\} = \{n/2, (n+2)/2\}$. Suppose, say, that $a_1 = n/2$ and $a_n = (n+2)/2$. Then, to achieve the maximum, we require that $\{a_3, a_5, \ldots, a_{n-1}\} = \{1, 2, \ldots, (n-2)/2\}$ and $\{a_2, a_4, \ldots, a_{n-2}\} = \{(n/2) + 2, (n/2) + 3, \ldots, n\}$. The maximum value of $(n^2 - 2)/2$ can be achieved with $2[(n-2)/2]!^2$ permutations.

2004:8. Let V be a complex n-dimensional inner product space. Prove that

$$|u|^2 |v|^2 - \frac{1}{4} |u - v|^2 |u + v|^2 \leq |(u, v)|^2 \leq |u|^2 |v|^2.$$

The solution to this problem appears in Chap. 7.

2005:2. Suppose that f is continuously differentiable on $[0, 1]$ and that $\int_0^1 f(x)dx = 0$. Prove that

$$2 \int_0^1 f(x)^2 dx \leq \int_0^1 |f'(x)| dx \cdot \int_0^1 |f(x)| dx.$$

The solution to this problem appears in Chap. 6.

2005:9 Let S be the set of all real-valued functions that are defined, positive and twice continuously differentiable on a neighbourhood of 0. Suppose that a and b are real parameters with $ab \neq 0$, $b < 0$. Define operators from S to $\mathbb{R}$ as follows:

$$A(f) = f(0) + af'(0) + bf''(0);$$

$$G(f) = \exp A(\log f).$$

(a) Prove that $A(f) \leq G(f)$ for $f \in S$;
(b) Prove that $G(f + g) \leq G(f) + G(g)$ for $f, g \in S$;
(c) Suppose that H is the set of functions in S for which $G(f) \leq f(0)$. Give examples of nonconstant functions, one in H and one not in H. Prove that, if $\lambda > 0$ and $f, g \in H$, then λf, $f + g$ and fg all belong to H.

The solution to this problem appears in Chap. 6.

2006:3. Let $p(x)$ be a polynomial of positive degree n with n distinct real roots $a_1 < a_2 < \cdots < a_n$. Let b be a real number for which $2b < a_1 + a_2$. Prove that

$$2^{n-1} |p(b)| \geq |p'(a_1)(b - a_1)|.$$

The solution to this problem appears in Chap. 2.

2009:7. Let $n \geq 2$. Minimize $a_1 + a_2 + \cdots + a_n$ subject to the constraints $0 \leq a_1 \leq a_2 \leq \cdots \leq a_n$ and $a_1 a_2 + a_2 a_3 + \cdots + a_{n-1} a_n + a_n a_1 = 1$. (When $n = 2$, the latter condition is $a_1 a_2 = 1$; when $n \geq 3$, the sum on the left has exactly n terms.)

Solution. If $n \geq 3$ and $a_i = 1/\sqrt{n}$ for each i, then $a_1 + a_2 + \cdots + a_n = \sqrt{n}$. If $a_1 = a_2 = \cdots = a_{n-2} = 0$ and $a_{n-1} = a_n = 1$, $a_1 + a_2 + \cdots + a_n = 2$. Therefore, when $n \geq 3$, the minimum does not exceed the lesser of $\sqrt{n}$ and 2.

Let $n = 2$. By the arithmetic-geometric means inequality, we have that

$$a_1 + a_2 \geq 2\sqrt{a_1 a_2}$$

so that the minimum is 2.

Let $n = 3$. Then

$$(a_1 + a_2 + a_3)^2 = a_1^2 + a_2^2 + a_3^2 + 2(a_1 a_2 + a_2 a_3 + a_3 a_1)$$

$$= \frac{1}{2}(a_1^2 + a_2^2) + \frac{1}{2}(a_2^2 + a_3^2) + \frac{1}{2}(a_3^2 + a_1^2) + 2$$

$$\geq a_1 a_2 + a_2 a_3 + a_3 a_1 + 2 = 3,$$

so that the minimum is $\sqrt{3}$.

Let $n = 4$. Then

$$(a_1 + a_2 + a_3 + a_4)^2 = a_1^2 + a_2^2 + a_3^2 + a_4^2 + 2 + 2a_1 a_3 + 2a_2 a_4$$

$$= (a_1 + a_3)^2 + (a_2 + a_4)^2 + 2$$

$$\geq 2[(a_1 + a_3)(a_2 + a_4)] + 2 = 4.$$

Therefore the minimum value of $a_1 + a_2 + a_3 + a_4$ is 2.

When $n \geq 5$, we have that

$$a_1 a_3 + a_2 a_4 + \cdots + a_{n-1} a_1 + a_n a_2$$

$$\geq a_1 a_2 + a_2 a_3 + \cdots + a_{n-1}[(a_1 - a_n) + a_n] + a_n a_1$$

$$= a_1 a_2 + a_2 a_3 + \cdots + a_{n-1} a_n + a_n a_1 + a_{n-1}(a_1 - a_n).$$

Hence

$$(a_1 + a_2 + \cdots + a_n)^2$$

$$= a_1^2 + a_2^2 + \cdots + a_{n-1}^2 + a_n^2 + 2(a_1 a_2 + a_2 a_3 + \cdots + a_{n-1} a_n + a_n a_1)$$

$$+ 2(a_1 a_3 + a_2 a_4 + \cdots + a_{n-1} a_1 + a_n a_2)$$

$$+ \sum \{a_i a_j : i - j \not\equiv \pm 1, \pm 2 \, (\mathrm{mod} \, n)\}$$

$$\geq a_1^2 + a_2^2 + \cdots + a_{n-1}^2 + a_n^2 + 4(a_1 a_2 + a_2 a_3 + \cdots + a_{n-1} a_n + a_n a_1)$$

$$+ 2a_{n-1}(a_1 - a_n)$$

$$= a_1^2 + a_2^2 + \cdots + a_{n-1}^2 + a_n^2 + 4 + [a_{n-1} + (a_1 - a_n)]^2$$
$$- a_{n-1}^2 - (a_1 - a_n)^2$$
$$= a_2^2 + \cdots + a_{n-2}^2 + 4 + (a_{n-1} + a_1 - a_n)^2 + 2a_1 a_n \geq 4.$$

Therefore the minimum value of $a_1 + a_2 + \cdots + a_n$ is equal to 2.

2011:2. Let u and v be positive reals. Minimize the larger of the two values

$$2u + \frac{1}{v^2} \quad \text{and} \quad 2v + \frac{1}{u^2}.$$

Solution 1. Observe that

$$\left(2u + \frac{1}{v^2} \right) + \left(2v + \frac{1}{u^2} \right) = \left(u + u + \frac{1}{u^2} \right) + \left(v + v + \frac{1}{v^2} \right)$$
$$\geq 3 \left(u \cdot u \cdot \frac{1}{u^2} \right)^{1/3} + 3 \left(v \cdot v \cdot \frac{1}{v^2} \right)^{1/3} = 6$$

by the arithmetic-geometric means inequality. It follows that one of $2u + v^{-2}$ and $2v + u^{-2}$ must be not less than 3. Since both assume the value 3 when $u = v = 1$, the required minimum value is 3.

Solution 2. Wolog, we may assume that $u \geq v$, whereupon

$$2u + \frac{1}{v^2} \geq 2v + \frac{1}{v^2} \geq 2v + \frac{1}{u^2},$$

with equality if and only if $u = v$. By checking the first derivative, it can be verified that $2v + v^{-2}$ achieves its minimum value of 3 when $v = 1$, so that $2u + v^{-2} \geq 3$ for all positive values of u and v, with equality if and only if $u = v = 1$.

2012:8. Determine the area of the set of points (x, y) in the plane that satisfy the two inequalities:

$$x^2 + y^2 \leq 2$$
$$x^4 + x^3 y^3 \leq xy + y^4.$$

The solution to this problem appears in Chap. 8.

2014:7. Suppose that $x_0, x_1, \cdots, x_n$ are real numbers. For $0 \leq i \leq n$, define

$$y_i = \max(x_0, x_1, \ldots, x_i).$$

Prove that

$$y_n^2 \leq 4x_n^2 - 4 \sum_{i=0}^{n-1} y_i (x_{i+1} - x_i).$$

When does equality occur?

Solution. We can simplify the problem. Suppose that for some $i < j$, $x_i = y_i = y_{i+1} = \cdots = y_j < y_{j+1}$. Then $y_{j+1} = x_{j+1} > x_i$ and

$$y_i(x_{i+1} - x_i) + y_{i+1}(x_{i+2} - x_{i+1}) + \cdots + y_j(x_{j+1} - x_j) = y_i(x_{j+1} - x_i),$$

so that all of $x_{i+1}, \cdots, x_j$ do not figure in the terms of the right side. Therefore, with no loss of generality, we can assume that

$$x_0 < x_1 < \cdots < x_{n-1}$$

so that $y_i = x_i$ for $0 \le i \le n - 1$, and y_n is equal to the maximum of x_{n-1} and x_n.

The right side of the inequality is equal to

$$4x_n^2 - 4[x_0(x_1 - x_0) + x_1(x_2 - x_1) + \cdots + x_{n-1}(x_n - x_{n-1})]$$
$$= 2x_0^2 + 2(x_1 - x_0)^2 + 2(x_2 - x_1)^2 + \cdots + 2(x_n - x_{n-1})^2 + 2x_n^2$$
$$= 2x_0^2 + 2(x_1 - x_0)^2 + 2(x_2 - x_1)^2 + \cdots + 2(x_{n-1} - x_{n-2})^2 + x_{n-1}^2$$
$$+ (2x_n - x_{n-1})^2.$$

Therefore, the right side is greater than or equal to both of $2x_n^2$ and x_{n-1}^2. The desired result follows.

Equality occurs if and only if $x_i = 0$ for each i.

Comment. It is natural to try a proof by induction. When $n = 1$, the right side is equal to

$$4x_1^2 - 4x_0x_1 + 4x_0^2 = 2x_1^2 + 2x_0^2 + 2(x_1 - x_0)^2$$

which is clearly greater than or equal to either of x_1^2 and x_0^2.

Suppose that the inequality holds for $n = m \ge 1$. Then

$$4x_{m+1}^2 - 4\sum_{i=0}^{m} y_i(x_{i+1} - x_i)$$

$$= 4x_{m+1}^2 - 4x_m^2 + \left[4x_m^2 - 4\sum_{i=0}^{m-1} y_i(x_{i+1} - x_i)\right] - 4y_m(x_{m+1} - x_m)$$

$$\ge 4x_{m+1}^2 - 4x_m^2 + y_m^2 - 4y_m(x_{m+1} - x_m) \equiv K.$$

Since y_{m+1} is the greater of x_{m+1} and y_m, there are two cases to consider.

When $y_{m+1} = x_{m+1}$, then, since $x_m \le y_m \le y_{m+1}$ and $2y_{m+1} - y_m \ge y_m$,

$$K = (2y_{m+1} - y_m)^2 + 4x_m(y_m - x_m) \ge (2y_{m+1} - y_m)^2 \ge y_{m+1}^2.$$

However, when $y_{m+1} = y_m$, then

$$K = y_m^2 + 4(x_{m+1} - x_m)(x_{m+1} + x_m - y_m)$$

and it is not clear whether the second term is always nonnegative. Can an induction argument be made to work?

[Problem **2014:7** is due to Mathias Beiglböck of the University of Vienna, and was communicated to the contest by Florian Herzig of the University of Toronto.]

2015:4. Determine all the values of the positive integer $n \geq 2$ for which the following statement is true, and for each, indicate when equality holds.

For any nonnegative real numbers $x_1, x_2, \ldots, x_n$,

$$(x_1 + x_2 + \cdots + x_n)^2 \geq n(x_1 x_2 + x_2 x_3 + \cdots + x_{n-1} x_n + x_n x_1),$$

where the right side has n summands.

Solution. Let $x_1 = x_2 = 1$ and $x_3 = x_4 = \cdots = x_n = 0$. Then the left side of the inequality is equal to 4 and the right side to n. Therefore a necessary condition for the inequality to hold for all sets of x_i is $n \leq 4$.

For $n = 2$, we find that

$$(x_1 + x_2)^2 - 2(x_1 x_2 + x_2 x_1) = (x_1 - x_2)^2 \geq 0,$$

so the inequality holds with equality if and only if $x_1 = x_2$.

For $n = 3$, we find that

$$2[(x_1 + x_2 + x_3)^2 - 3(x_1 x_2 + x_2 x_3 + x_3 x_1)] = (x_1 - x_2)^2 + (x_2 - x_3)^2 + (x_3 - x_1)^2 \geq 0,$$

so the inequality holds with equality if and only if $x_1 = x_2 = x_3$.

For $n = 4$, we find that

$$\begin{aligned}
(x_1 + x_2 &+ x_3 + x_4)^2 - 4(x_1 x_2 + x_2 x_3 + x_3 x_4 + x_4 x_1) \\
&= x_1^2 + x_2^2 + x_3^2 + x_4^2 + 2x_1 x_3 + 2x_2 x_4 - 2x_1 x_2 \\
&\quad - 2x_2 x_3 - 2x_3 x_4 - 2x_4 x_1 \\
&= (x_1 - x_2)^2 + (x_3 - x_4)^2 + 2(x_1 - x_2)(x_3 - x_4) \\
&= (x_1 - x_2 + x_3 - x_4)^2 \geq 0,
\end{aligned}$$

so the inequality holds with equality if and only if $x_1 + x_3 = x_2 + x_4$.

Comment. Another case that might be tried is $x_i = i - 1$ for $1 \leq i \leq n$. Then $x_1 + \cdots + x_n = \frac{1}{2}n(n-1)$ and $x_1 x_2 + \cdots + x_{n-1} x_n + x_n x_1 = \frac{1}{3}n(n-1)(n-2)$. The left side minus the right is equal to $\frac{1}{12}n^2(n-1)(5-n)$, which establishes that $n \leq 5$. In fact, in the $n = 5$ case, equality occurs whenever the x_i form an arithmetic progression, so a counterexample has to be found elsewhere.

CHAPTER 4

Sequences and Series

2001:6. Prove that, for each positive integer n, the series

$$\sum_{k=1}^{\infty} \frac{k^n}{2^k}$$

converges to twice an odd integer not less than $(n+1)!$.

Solution 1. Convergence of the series results from either the ratio or the root test. For nonnegative integers n, let

$$S_n = \sum_{k=1}^{\infty} \frac{k^n}{2^k}.$$

Then $S_0 = 1$ and

$$
\begin{aligned}
S_n - \frac{1}{2}S_n &= \sum_{k=1}^{\infty} \frac{k^n}{2^k} - \sum_{k=1}^{\infty} \frac{k^n}{2^{k+1}} \\
&= \sum_{k=1}^{\infty} \frac{k^n}{2^k} - \sum_{k=1}^{\infty} \frac{(k-1)^n}{2^k} \\
&= \sum_{k=1}^{\infty} \frac{k^n - (k-1)^n}{2^k} \\
&= \sum_{k=1}^{\infty} \left[\binom{n}{1}\frac{k^{n-1}}{2^k} - \binom{n}{2}\frac{k^{n-2}}{2^k} + \binom{n}{3}\frac{k^{n-3}}{2^k} - \cdots + (-1)^{n-1}\frac{1}{2^k} \right]
\end{aligned}
$$

whence

$$S_n = 2\left[\binom{n}{1}S_{n-1} - \binom{n}{2}S_{n-2} + \binom{n}{3}S_{n-3} - \cdots + (-1)^{n-1} \right].$$

An induction argument establishes that S_n is twice an odd integer.

© Springer International Publishing Switzerland 2016
E.J. Barbeau, *University of Toronto Mathematics Competition (2001–2015)*, Problem Books in Mathematics,
DOI 10.1007/978-3-319-28106-3_4

Observe that $S_0 = 1$, $S_1 = 2$, $S_2 = 6$ and $S_3 = 26$. We prove by induction that, for each $n \geq 0$,

$$S_{n+1} \geq (n+2)S_n$$

from which the desired result will follow. Suppose that we have established this for $n = m - 1$. Now

$$S_{m+1} = 2\left[\binom{m+1}{1}S_m - \binom{m+1}{2}S_{m-1} + \binom{m+1}{3}S_{m-2}\right.$$

$$\left. - \binom{m+1}{4}S_{m-3} + \cdots\right].$$

For each positive integer r not exceeding $\frac{1}{2}(m+1)$,

$$\binom{m+1}{2r-1}S_{m-2r+2} - \binom{m+1}{2r}S_{m-2r+1}$$

$$\geq \left[\binom{m+1}{2r-1}(m-2r+3) - \binom{m+1}{2r}\right]S_{m-2r+1}$$

$$= \binom{m+1}{2r-1}\left[(m-2r+3) - \left(\frac{m-2r+2}{2r}\right)\right]S_{m-2r+1} \geq 0 \, .$$

When $r = 1$, we get inside the square brackets the quantity

$$(m+1) - \frac{m}{2} = \frac{m+2}{2}$$

while when $r > 1$, we get

$$(m - 2r + 3) - \left(\frac{m-2r+2}{2r}\right) > (m-2r+3) - (m-2r+2) = 1.$$

Hence

$$S_{m+1} \geq 2\left[\binom{m+1}{1}S_m - \binom{m+1}{2}S_{m-1}\right]$$

$$\geq 2\left[(m+1)S_m - \frac{m(m+1)}{2} \cdot \frac{1}{m+1}S_m\right]$$

$$= 2\left[m+1 - \frac{m}{2}\right]S_m = (m+2)S_m.$$

Solution 2. Define S_n as in the foregoing solution. Then, for $n \geq 1$,

$$S_n = \frac{1}{2} + \sum_{k=2}^{\infty} \frac{k^n}{2^k}$$

$$= \frac{1}{2} + \frac{1}{2} \sum_{k=1}^{\infty} \frac{(k+1)^n}{2^k}$$

$$= \frac{1}{2} + \frac{1}{2} \sum_{k=1}^{\infty} \frac{k^n + \binom{n}{1} k^{n-1} + \cdots + \binom{n}{n-1} k + 1}{2^k}$$

$$= \frac{1}{2} + \frac{1}{2} \left[S_n + \binom{n}{1} S_{n-1} + \cdots + \binom{n}{n-1} S_1 + 1 \right]$$

whence $S_1 = 2$ and

$$S_n = \binom{n}{1} S_{n-1} + \binom{n}{2} S_{n-2} + \cdots + \binom{n}{n-1} S_1 + 2$$

for $n \geq 2$. It is easily checked that $S_k \equiv 2 \pmod 4$ for $k = 1$. As an induction hypothesis, suppose this holds for $1 \leq k \leq n - 1$. Then, modulo 4, the right side is congruent to

$$2 \left[\sum_{k=0}^{n} \binom{n}{k} - 2 \right] + 2 = 2(2^n - 2) + 2 = 2^{n+1} - 2,$$

and the desired result follows.

For $n \geq 1$,

$$\frac{S_{n+1}}{S_n} = \frac{\binom{n+1}{1} S_n + \binom{n+1}{2} S_{n-1} + \cdots + \binom{n+1}{n} S_1 + 2}{S_n}$$

$$= (n+1) + \frac{\binom{n+1}{2} S_{n-1} + \binom{n+1}{3} S_{n-2} + \cdots + (n+1) S_1 + 2}{\binom{n}{1} S_{n-1} + \binom{n}{2} S_{n-2} + \cdots + n S_1 + 2}$$

$$\geq (n+1) + 1 = n + 2,$$

since each term in the numerator of the latter fraction exceeds each corresponding term in the denominator.

Solution 3 (of the first part). [P. Gyrya] Let $f(x)$ be a differentiable function and let D be the differentiation operator. Define the operator L by

$$L(f)(x) = x \cdot D(f)(x).$$

Suppose that $f(x) = (1 - x)^{-1} = \sum_{k=0}^{\infty} x^k$. Then, it is standard that $L^n(f)(x)$ has a power series expansion obtained by term-by-term differentiation that converges absolutely for $|x| < 1$. By induction, it can be shown that the series given in the problem is, for each nonnegative integer n, $L^n(f)(1/2)$.

It is straightforward to verify that

$$L((1-x)^{-1}) = x(1-x)^{-2}$$
$$L^2((1-x)^{-1}) = x(1+x)(1-x)^{-3}$$
$$L^3((1-x)^{-1}) = x(1+4x+x^2)(1-x)^{-4}$$
$$L^4((1-x)^{-1}) = x(1+11x+11x^2+x^3)(1-x)^{-5}.$$

In general, a straightforward induction argument yields that for each positive integer n,

$$L^n(f)(x) = x(1+a_{n,1}x+\cdots+a_{n,n-2}x^{n-2}+x^{n-1})(1-x)^{-(n+1)}$$

for some integers $a_{n,1},\cdots,a_{n,n-2}$. Hence

$$L^n(f)(1/2) = 2(2^{n-1}+a_{n,1}2^{n-2}+\cdots+a_{n,n-2}2+1),$$

yielding the desired result.

2003:1. Evaluate

$$\sum_{n=1}^{\infty} \arctan\left(\frac{2}{n^2}\right).$$

Solution 1. Let $a_n = \arctan n$ for $n \geq 0$. Then

$$\tan(a_{n+1}-a_{n-1}) = \frac{(n+1)-(n-1)}{1+(n^2-1)} = \frac{2}{n^2}$$

for $n \geq 1$. Then

$$\sum_{n=1}^{m} \arctan\frac{2}{n^2} = \arctan(m+1) + \arctan m - \arctan 1 - \arctan 0.$$

Letting $m \to \infty$ yields the answer $\pi/2 + \pi/2 - \pi/4 - 0 = 3\pi/4$.

Solution 2. Let $b_n = \arctan(1/n)$ for $n \geq 0$. Then

$$\tan(b_{n-1}-b_{n+1}) = \frac{2}{n^2}$$

for $n \geq 2$, whence

$$\sum_{n=1}^{m} \arctan\frac{2}{n^2} = \arctan 2 + \sum_{n=2}^{m}(b_{n-1}-b_{n+1})$$

$$= \arctan 2 + \arctan 1 + \arctan\frac{1}{2} - \arctan\frac{1}{m} - \arctan\frac{1}{m+1}$$

$$= (\arctan 2 + \text{arccot } 2) + \arctan 1 - \arctan\frac{1}{m} - \arctan\frac{1}{m+1}$$

$$= \frac{\pi}{2} + \frac{\pi}{4} - \arctan\frac{1}{m} - \arctan\frac{1}{m+1}$$

for $m \geq 3$, from which the result follows by letting m tend to infinity.

Solution 3. [S. Huang] Let $s_n = \sum_{k=1}^{n}\arctan(2/k^2)$ and $t_n = \tan s_n$. Then $\{t_n\} = \{2, \infty, -9/2, -14/5, -20/9, \dots\}$ where the numerators of the fractions are $\{-2, -5, -9, -14, -20, \dots\}$ and the denominators are $\{-1, 0, 2, 5, 8, \dots\}$. We conjecture that

$$t_n = \frac{-n(n+3)}{(n-2)(n+1)}$$

for $n \geq 1$. This is true for $1 \leq n \leq 5$. Suppose that it holds to $n = k - 1 \geq 5$, so that $t_{k-1} = -(k-1)(k+2)/(k-3)k$. Then

$$t_k = \frac{t_{k-1} + 2k^{-2}}{1 - 2t_{k-1}k^{-2}}$$

$$= \frac{-k^2(k-1)(k+2) + 2(k-3)k}{k^3(k-3) + 2(k-1)(k+2)}$$

$$= \frac{-k(k+3)(k^2 - 2k + 2)}{(k-2)(k+1)(k^2 - 2k + 2)} = \frac{-k(k+3)}{(k-2)(k+1)}.$$

The desired expression for t_n holds by induction and so $\lim_{n \to \infty} t_n = -1$. For $n \geq 3$, $t_n < 0$ and $\tan^{-1}(2/n^2) < \pi/2$, so we must have $\pi/2 < s_n < \pi$ and $s_n = \pi - \tan^{-1} t_n$. Therefore

$$\lim_{n \to \infty} s_n = \arctan(\pi + \lim_{n \to \infty} t_n) = \pi - (\pi/4) = (3\pi)/4.$$

2003:4. Show that the positive integer n divides the integer nearest to

$$\frac{(n+1)!}{e}.$$

Solution. By Taylor's Theorem, we have that

$$e^{-1} = 1 - \frac{1}{1!} + \frac{1}{2!} - \frac{1}{3!} + \cdots + (-1)^{n+1}\frac{1}{(n+1)!} + (-1)^{n+2}\frac{e^c}{(n+2)!}$$

where $-1 < c < 0$. Hence

$$(n+1)!e^{-1} = (n+1)! - (n+1)! + \frac{(n+1)!}{2!} - \frac{(n+1)!}{3!}$$

$$+ \cdots + (-1)^n[(n+1) - 1] + (-1)^{n-2}\frac{e^c}{n+2}.$$

The last term does not exceed $1/(n+2)$, which is less than $1/2$. The second last term is equal to $\pm n$, and each previous term has a factor n in the numerator that is not cancelled out by the denominator. Since the sum of all but the last term is an integer divisible by n and is the nearest integer to $(n+1)!/e$, the result holds.

2004:2. Prove that

$$\int_0^1 x^x \, dx = 1 - \frac{1}{2^2} + \frac{1}{3^3} - \frac{1}{4^4} + \frac{1}{5^5} + \cdots.$$

Solution. First, let

$$I(m, n) = \int_0^1 x^m (\log x)^n \, dx$$

for nonnegative integers m and n. Then $I(0,0) = 1$ and $I(m,0) = 1/(m+1)$ for every nonnegative integer m. Taking the parts $u = (\log x)^n$, $dv = x^m dx$ and noting that $\lim_{x \downarrow 0} x^{m+1} (\log x)^n = 0$, we find that $I(m,n) = -(n/(m+1)) I(m, n-1)$ for $n \geq 1$, whence

$$I(m,n) = \frac{(-1)^n n!}{(m+1)^{n+1}}$$

for each nonnegative integer n. In particular,

$$I(k,k) = \frac{(-1)^k k!}{(k+1)^{(k+1)}}$$

for each nonnegative integer k.

Using the fact that the series is uniformly convergent and term-by-term integration is possible, we find that

$$\int_0^1 x^x \, dx = \int_0^1 e^{x \log x} \, dx = \sum_{k=0}^{\infty} \int_0^1 \frac{(x \log x)^k}{k!} \, dx$$

$$= 1 + \sum_{k=1}^{\infty} \frac{(-1)^k}{(k+1)^{(k+1)}} = 1 - \frac{1}{2^2} + \frac{1}{3^3} - \cdots .$$

2007:6. Let $h(n)$ denote the number of finite sequences $\{a_1, a_2, \ldots, a_k\}$ of positive integers exceeding 1 for which $k \geq 1$, $a_1 \geq a_2 \geq \cdots \geq a_k$ and $n = a_1 a_2 \cdots a_k$. (For example, if $n = 20$, there are four such sequences $\{20\}$, $\{10, 2\}$, $\{5, 4\}$ and $\{5, 2, 2\}$ and $h(20) = 4$.)

Prove that

$$\sum_{n=1}^{\infty} \frac{h(n)}{n^2} = 1.$$

Solution 1. We have that

$$1 + \sum_{n=2}^{\infty} \frac{h(n)}{n^2} = \prod_{r=2}^{\infty} \left(\sum_{k=0}^{\infty} \frac{1}{r^{2k}} \right)$$

$$= \prod_{r=2}^{\infty} \frac{1}{1 - 1/r^2} = \prod_{r=2}^{\infty} \frac{r^2}{r^2 - 1}$$

$$= \lim_{n \to \infty} \prod_{r=2}^{n} \left(\frac{r}{r-1} \right) \left(\frac{r}{r+1} \right) = \lim_{n \to \infty} \frac{2n}{n+1} = 2,$$

from which the result follows.

Solution 2. [J. Kramar] Observe that

$$\sum_{n=1}^{\infty} \frac{h(n)}{n^2} = \sum \{(a_1 a_2 \cdots a_k)^{-2} : a_1 \geq a_2 \geq \cdots \geq a_k \geq 2\}.$$

For $m \geq 2$, let

$$b_m = \sum \{(a_1 a_2 \cdots a_k)^{-2} : a_1 = m \geq a_2 \geq \cdots \geq a_k \geq 2\}.$$

Since

$$b_m = \frac{1}{m^2} + \frac{1}{m^2}(b_m + b_{m-1} + \cdots + b_2),$$

then, for $m \geq 3$,

$$(m^2 - 1)b_m = 1 + b_2 + \cdots + b_{m-1}.$$

Note that

$$b_2 = \frac{1}{2^2} + \frac{1}{4^2} + \frac{1}{8^2} + \cdots = \frac{1}{3}.$$

Assume as an induction hypothesis that, for $2 \leq k \leq m - 1$, $b_k = 2/(k(k+1))$. Then

$$(m+1)(m-1)b_m = 1 + b_2 + \cdots + b_{m-1} = 1 + 2 \sum_{k=2}^{m-1} \frac{1}{k(k+1)}$$

$$= 1 + 2 \left(\sum_{k=2}^{m-1} \frac{1}{k} - \frac{1}{k+1} \right) = 1 + 2 \left(\frac{1}{2} - \frac{1}{m} \right)$$

$$= 2 - \frac{2}{m} = 2 \left(\frac{m-1}{m} \right),$$

so that $b_m = 2/(m(m+1))$.

Hence

$$\sum_{n=1}^{\infty} \frac{h(n)}{n^2} = \sum_{m=2}^{\infty} b_m = 2 \sum_{m=2}^{\infty} \left(\frac{1}{m} - \frac{1}{m+1} \right) = 1.$$

2008:3. Suppose that a is a real number and the sequence $\{a_n\}$ is defined recursively by $a_0 = a$ and

$$a_{n+1} = a_n(a_n - 1)$$

for $n \geq 0$. Find the values of a for which the sequence $\{a_n\}$ converges.

Solution. When the sequence converges, the limit b must satisfy the equation $b = b(b-1)$ so that $b = 0$ or $b = 2$. It is clear that the sequence converges when $a = -1, 0, 1, 2$. If $a > 2$, then an induction argument shows that $\{a_n\}$ is increasing and unbounded. If $a < -1$, then $a_1 > 2$ and the sequence again diverges.

If $-1 < a < 2$, we show that the sequence will have an entry in the interval $[0, 1]$. Suppose that $1 < a < 2$, then $a(a - 1) < a$, so that a_n will initially decrease until it arrives in the interval $(0, 1]$. If $-1 < a < 0$, then $a_1 = (-a)(1 - a)$ will lie in $(0, 2)$.

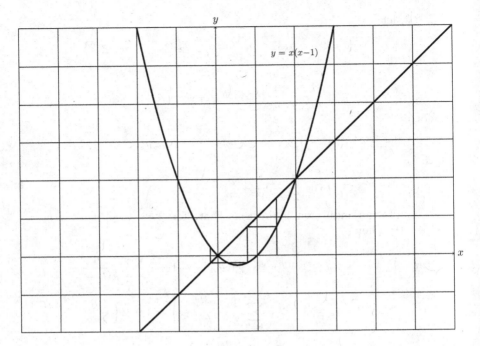

Fig. 4.1. A cobweb diagram

It remains to analyze the situation that $0 < a < 1$. Then $-1/4 \le a_1 < 0$ and $0 < a_2 \le 5/16 < 1$. Thus, the sequence alternates between the intervals $[-1/4, 0)$ and $(0, 1)$. Observe that

$$a_{n+2} = a_{n+1}(a_{n+1} - 1) = a_n(a_n - 1)(a_n^2 - a_n - 1)$$
$$= a_n(1 - a_n^2(2 - a_n)).$$

When n is even, then a_n and a_{n+2} are both positive and $a_{n+2} < a_n$. When n is odd, then a_n and a_{n+2} are both negative and $1 - a_n^2(2 - a_n) \in (0, 1)$, so that the sequence $\{a_{2m}\}$ decreases to a limit u and $\{a_{2m+1}\}$ increases to a limit v. We must have that $u = v(v - 1)$ and $v = u(u - 1)$ so that $u = u(u - 1)(u^2 - u - 1)$ or $u^3(2 - u) = 0$. Since $u \ne 2$, we must have $u = v = 0$.

Hence the sequence converges if and only if $-1 \le a \le 2$; the limit is 2 when $a = -1, 2$ and 0 when $-1 < a < 2$.

Comment. The situation can also be analyzed using a cobweb diagram on the graph of the function $y = x(x - 1)$. On the same axes, plot $y = x$ and $y = f(x)$, and then follow the progress of the points $(a_1, a_2) = (a_1, f(a_1))$, (a_2, a_2), $(a_2, a_3) = (a_2, f(a_2))$, (a_3, a_3), and so on. The positions of the abscissae will indicate the behaviour of a_n as n increases. See Fig. 4.1.

2008:9. For each positive integer n, let

$$S_n = \sum_{k=1}^{n} \frac{2^k}{k^2}.$$

Prove that S_{n+1}/S_n is not a rational function of n.

The solution to this problem appears in Chap. 2.

2009:4. Let $\{a_n\}$ be a real sequence for which

$$\sum_{n=1}^{\infty} \frac{a_n}{n}$$

converges. Prove that

$$\lim_{n\to\infty} \frac{a_1 + a_2 + \cdots + a_n}{n} = 0.$$

Solution. Recall Abel's Partial Summation Formula:

$$\sum_{k=1}^{n} x_k y_k = (x_1 + x_2 + \cdots + x_n)y_{n+1} + \sum_{k=1}^{n} (x_1 + x_2 + \cdots + x_k)(y_k - y_{k+1}).$$

Applying this to $x_n = a_n/n$ and $y_n = n$. we obtain that

$$a_1 + a_2 + \cdots + a_n = (x_1 + x_2 + \cdots + x_n)(n+1) - \sum_{k=1}^{n}(x_1 + x_2 + \cdots + x_k),$$

whence

$$\frac{a_1 + a_2 + \cdots + a_n}{n} = \left(\sum_{k=1}^{n} x_k\right)\left(1 + \frac{1}{n}\right) - \frac{1}{n}\sum_{k=1}^{n}(x_1 + x_2 + \cdots + x_k).$$

Let

$$L = \lim_{n\to\infty} \sum_{k=1}^{n} x_k = \lim_{n\to\infty} \sum_{k=1}^{n} \left(\frac{a_k}{k}\right).$$

Then

$$L = \lim_{n\to\infty} \left(1 + \frac{1}{n}\right)\left(\sum_{k=1}^{n} x_k\right) = \lim_{n\to\infty} \frac{1}{n}\sum_{k=1}^{n}(x_1 + x_2 + \cdots + x_k).$$

To see the latter equality, suppose that $u_n = x_1 + x_2 + \cdots + x_n$ and $v_n = (1/n)(u_1 + u_2 + \cdots + u_n)$, so that

$$v_n - u_n = \frac{1}{n}[(u_1 - u_n) + (u_2 - u_n) + \cdots (u_n - u_n)].$$

Let $\epsilon > 0$ be given. First select n_1 so that $|u_k - u_n| < \frac{1}{2}\epsilon$ for $n \geq k \geq n_1$. Since $\lim u_n = L$, there exists a positive integer M for which $|u_n| < M$ for each n. Select $n_2 > n_1$ such that, for all $n > n_2$, $2Mn_1/n < \epsilon/2$. Then, for all $n > n_2$,

$$|v_n - u_n| \leq \frac{1}{n}\left[\sum_{k=1}^{n_1} |u_k - u_n| + \sum_{k=n_1+1}^{n} |u_k - u_n|\right]$$

$$\leq \frac{1}{n}\left(\frac{2Mn_1}{n}\right) + \frac{n - n_1}{n}\left(\frac{\epsilon}{2}\right) < \left(\frac{\epsilon}{2} + \frac{\epsilon}{2}\right) = \epsilon.$$

It follows that $\lim v_n = L$, and the desired result obtains.

2010:2. Let $u_0 = 1$, $u_1 = 2$ and $u_{n+1} = 2u_n + u_{n-1}$ for $n \geq 1$. Prove that, for every nonnegative integer n,

$$u_n = \sum \left\{\frac{(i + j + k)!}{i!j!k!} : i, j, k \geq 0, i + j + 2k = n\right\}.$$

The solution to this problem appears in Chap. 10.

2011:4. Let $\{b_n : n \geq 1\}$ be a sequence of positive real numbers such that

$$3b_{n+2} \geq b_{n+1} + 2b_n$$

for every positive integer n. Prove that either the sequence converges or that it diverges to infinity.

Solution 1. For $n \geq 1$, let $c_n = \min(b_n, b_{n+1})$. Then $b_{n+1} \geq c_n$ and $b_{n+2} \geq \frac{1}{3}b_{n+1} + \frac{2}{3}b_n \geq c_n$, so that $c_{n+1} \geq c_n$ for $n \geq 1$.

Since $b_n \geq c_n$ for all n, $\{b_n\}$ diverges to infinity if $\{c_n\}$ does.

Otherwise, let $\{c_n\}$ be bounded. Since the sequence increases, it increases to a finite limit m. Let $\epsilon > 0$ be given. Suppose that N is selected so that $m - \frac{1}{2}\epsilon < c_n \leq m$ for $n \geq N$. If b_n is equal to either c_n or c_{n-1}, then $m - \epsilon < b_n < m$ for $n > N$. If b_n is equal to neither c_{n-1} nor c_n, then $c_{n-1} = b_{n-1}$ and $c_n = b_{n+1}$, whence

$$b_n \leq 3b_{n+1} - 2b_{n-1} = 3c_n - 2c_{n-1} \leq 3m - 2(m - \epsilon/2) = m + \epsilon.$$

Thus, $m - \epsilon < b_n < m + \epsilon$ for $n > N$. It follows that $\lim_{n\to\infty} b_n = m$.

Solution 2. [A. Remorov] Let $u = \liminf_{n\to\infty} b_n$ and $v = \limsup_{n\to\infty} b_n$. If $\{b_n\}$ is unbounded, then, given $M > 0$, there exists r such that $b_r > 3M$. Then $b_{r+1} > \frac{1}{3}b_r > M$ and $b_{r+2} > \frac{2}{3}b_r > M$, whereupon it can be established by induction that $b_n > M$ for $n > r$. Hence $u \geq M$ for each $M > 0$ and $\lim_{n\to\infty} b_n = +\infty$.

Otherwise, $\{b_n\}$ is bounded and v is finite. Suppose, if possible, that $v > u$. Let s be selected so that $b_n > u - \frac{1}{3}(v - u)$ for $n \geq s$. If it happens that $b_{n+1} > u + \frac{3}{4}(v - u)$ for some $n \geq s$, then

$$b_{n+2} > \frac{u}{3} + \frac{1}{4}(v - u) + \frac{2u}{3} - \frac{1}{6}(v - u) = u + \frac{1}{12}(v - u),$$

from which it follows that $b_{n+k} > u + \frac{1}{12}(v - u)$ for all $k > 0$. But this would entail that $u \geq u + \frac{1}{12}(v - u)$, which is false. Therefore $b_{n+1} < u + \frac{3}{4}(v - u)$ for all $n \geq s$. Hence $v \leq u + \frac{3}{4}(v - u)$, or $v - u < \frac{3}{4}(v - u)$, which is false. Therefore $v = u$ and the result follows.

2013:10.

(a) Let f be a real-valued function defined on the real number field $\mathbb{R}$ for which $|f(x) - f(y)| < |x - y|$ for any pair (x, y) of distinct elements of $\mathbb{R}$. Let $f^{(n)}$ denote the nth composite of f defined by $f^{(1)}(x) = f(x)$ and $f^{(n+1)}(x) = f(f^{(n)}(x))$ for $n \geq 2$. Prove that exactly one of the following situations must occur:

(i) $\lim_{n \to +\infty} f^{(n)}(x) = +\infty$ for each real x;

(ii) $\lim_{n \to +\infty} f^{(n)}(x) = -\infty$ for each real x;

(iii) there is a real number z such that

$$\lim_{n \to +\infty} f^{(n)}(x) = z$$

for each real x.

(b) Give examples to show that each of the three cases in (a) can occur.

The solution to this problem appears in Chap. 6.

2014:3. Let n be a positive integer. A finite sequence $\{a_1, a_2, \ldots, a_n\}$ of positive integers a_i is said to be *tight* if and only if $1 \leq a_1 < a_2 < \cdots < a_n$, all $\binom{n}{2}$ differences $a_j - a_i$ with $i < j$ are distinct, and a_n is as small as possible.

(a) Determine a tight sequence for $n = 5$.

(b) Prove that there is a polynomial $p(n)$ of degree not exceeding 3 such that $a_n \leq p(n)$ for every tight sequence $\{a_i\}$ with n entries.

The solution to this problem appears in Chap. 11.

2014:5. Let n be a positive integer. Prove that

$$\sum_{k=1}^{n} \frac{1}{k\binom{n}{k}} = \sum_{k=1}^{n} \frac{1}{k2^{n-k}} = \frac{1}{2^{n-1}} \sum_{k=1}^{n} \frac{2^{k-1}}{k} = \frac{1}{2^{n-1}} \sum \left\{ \frac{\binom{n}{k}}{k} : k \text{ odd}, 1 \leq k \leq n \right\}.$$

Solution. The values of each of the sums for $1 \leq n \leq 5$ are $1, 1, 5/6, 2/3, 8/15$.

The equality of the second and third sums is straightforward to establish. Let S_n denote the leftmost sum. Since $k\binom{n}{k} = n\binom{n-1}{k-1}$, we have that

$$S_n = \sum_{k=1}^{n} \left[n\binom{n-1}{k-1} \right]^{-1}.$$

Using the fact that

$$\left[(n+1)\binom{n}{k-1} \right]^{-1} + \left[(n+1)\binom{n}{k} \right]^{-1} = \left[n\binom{n-1}{k-1} \right]^{-1},$$

for $1 \leq k \leq n$, it can be shown that $S_1 = 1$ and that

$$S_{n+1} = \frac{S_n}{2} + \frac{1}{n+1}$$

for $n \geq 1$.

Therefore

$$S_n = \frac{1}{2^{n-1}} + \frac{1}{2 \cdot 2^{n-2}} + \frac{1}{3 \cdot 2^{n-3}} + \cdots + \frac{1}{k \cdot 2^{n-k}} + \cdots + \frac{1}{n}$$

$$= \sum_{k=1}^{n} \frac{1}{k \cdot 2^{n-k}} = \frac{1}{2^{n-1}} \sum_{k=1}^{n} \frac{2^{k-1}}{k}.$$

For each positive integer n, let

$$T_n = \frac{1}{2^{n-1}} \sum_{k \text{ odd}} \frac{\binom{n}{k}}{k}.$$

Then $T_1 = 1$ and

$$2T_{n+1} - T_n = \frac{1}{2^{n-1}} \sum_{k \text{ odd}} \frac{\binom{n+1}{k} - \binom{n}{k}}{k} = \frac{1}{2^{n-1}} \sum_{k \text{ odd}} \frac{\binom{n}{k-1}}{k}$$

$$= \frac{1}{(n+1)2^{n-1}} \sum_{k \text{ odd}} \binom{n+1}{k} = \frac{2^n}{(n+1)2^{n-1}} = \frac{2}{n+1}.$$

Therefore the recursions for S_n and T_n agree when $n = 1$ and satisfy the same equations. Thus the result holds.

2014:9. Let $\{a_n\}$ and $\{b_n\}$ be positive real sequences such that

$$\lim_{n \to \infty} \frac{a_n}{n} = u > 0$$

and

$$\lim_{n \to \infty} \left(\frac{b_n}{a_n} \right)^n = v > 0.$$

Prove that

$$\lim_{n \to \infty} \left(\frac{b_n}{a_n} \right) = 1$$

and

$$\lim_{n \to \infty} (b_n - a_n) = u \log v.$$

Solution. Observe that

$$\lim_{n \to \infty} \frac{b_n}{a_n} = \lim_{n \to \infty} \exp \frac{1}{n} \log \left(\frac{b_n}{a_n} \right)^n = \exp(0 \cdot v) = \exp 0 = 1.$$

Suppose that $v \neq 1$. Then for sufficiently large values of n, $b_n \neq a_n$ and we have that

$$(b_n - a_n)\left[\left(\frac{a_n}{b_n - a_n}\right)\log\left(1 + \frac{b_n - a_n}{a_n}\right)\right] = \frac{a_n}{n}\log\left(\frac{b_n}{a_n}\right)^n.$$

Note that

$$\lim_{n\to\infty}\frac{b_n - a_n}{a_n} = \lim_{n\to\infty}\left(\frac{b_n}{a_n} - 1\right) = 0$$

and that $\lim_{t\downarrow 0} t^{-1}\log(1+t) = 1$, so that

$$\left[\lim_{n\to\infty}(b_n - a_n)\right]\cdot 1 = u\log v$$

as desired.

If $v = 1$, we need to exercise more care, as b_n could equal a_n infinitely often. In this case, the indices can be partitioned into two sets A and B, where respectively $b_n = a_n$ for $n \in A$ and $b_n \neq a_n$ for $n \in B$; either of these sets could be infinite or finite. If infinite, taking the limit over $n \in A$ and over $n \in B$ yields the result $0 = u\log v$, and the problem is solved.

[Problem **2014:9** was contributed by the AN-anduud Problem Solving Group in Ulaanbataar, Mongolia.]

2015:8. Let $\{a_n\}$ and $\{b_n\}$ be two *decreasing* positive real sequences for which

$$\sum_{n=1}^{\infty} a_n = \infty$$

and

$$\sum_{n=1}^{\infty} b_n = \infty.$$

Let I be a subset of the natural numbers, and define the sequence $\{c_n\}$ by

$$c_n = \begin{cases} a_n, & \text{if } n \in I \\ b_n, & \text{if } n \notin I. \end{cases}$$

Is it possible for $\sum_{n=1}^{\infty} c_n$ to converge?

Solution. The answer is *yes*. Let $s_{-1} = 0$, and for $n \geq 0$, let

$$s_n = \sum_{k=0}^{n} 2^{2^k}.$$

For $n \geq 0$, let I_n be the set of positive integers k for which $s_{n-1} < k \leq s_n$. Define

$$a_k = 2^{-2^{2n}-n} \quad \text{and} \quad b_k = 2^{-2^{2n}} \quad \text{when} \quad k \in I_{2n};$$

$$a_k = 2^{-2^{2n+1}} \quad \text{and} \quad b_k = 2^{-2^{2n+1}-n} \quad \text{when} \quad k \in I_{2n+1}.$$

It can be verified that both $\{a_n\}$ and $\{b_n\}$ are decreasing.

Let

$$I = \bigcup_{n=0}^{\infty} I_{2n}.$$

Then, for each nonnegative n,

$$\sum_{k \in I_{2n+1}} a_k = \sum_{k \in I_{2n}} b_k = 1.$$

Therefore

$$\sum_{n \in I} a_n = 2;$$

$$\sum_{n \notin I} a_n = \infty;$$

$$\sum_{n \in I} b_n = \infty;$$

and

$$\sum_{n \notin I} b_n = 2.$$

[Problem **2015:8** was contributed by Franklin Vera Pacheco.]

2015:10. (a) Let

$$g(x, y) = x^2 y + x y^2 + xy + x + y + 1.$$

We form a sequence $\{x_0\}$ as follows: $x_0 = 0$. The next term x_1 is the unique solution -1 of the linear equation $g(t, 0) = 0$. For each $n \geq 2$, x_n is the solution other than x_{n-2} of the equation $g(t, x_{n-1}) = 0$.

Let $\{f_n\}$ be the Fibonacci sequence determined by $f_0 = 0$, $f_1 = 1$ and $f_n = f_{n-1} + f_{n-2}$ for $n \geq 2$. Prove that, for any nonnegative integer k,

$$x_{2k} = \frac{f_k}{f_{k+1}} \quad \text{and} \quad x_{2k+1} = -\frac{f_{k+2}}{f_{k+1}}.$$

(b) Let

$$h(x, y) = x^2 y + x y^2 + \beta xy + \gamma(x + y) + \delta$$

be a polynomial with real coefficients β, γ, δ. We form a bilateral sequence $\{x_n : n \in \mathbb{Z}\}$ as follows. Let $x_0 \neq 0$ be given arbitrarily. We select x_{-1} and x_1 to be the two solutions of the quadratic equation $h(t, x_0) = 0$ in either order. From here, we can define inductively the terms of the sequence for positive and negative values of the index so that x_{n-1} and x_{n+1} are the two solutions of the equation $h(t, x_n) = 0$. We suppose that at each stage, neither of these solutions is zero.

Prove that the sequence $\{x_n\}$ has period 5 (i.e. $x_{n+5} = x_n$ for each index n) if and only if $\gamma^3 + \delta^2 - \beta\gamma\delta = 0$.

The solution to this problem appears in Chap. 2.

CHAPTER 5

Calculus and its Applications

2001:2. Let $O = (0,0)$ and $Q = (1,0)$. Find the point P on the line with equation $y = x + 1$ for which the angle OPQ is a maximum.

Solution 1. For the point P to maximize $\angle OPQ$, it must be a point of tangency of a circle with chord OQ and the line of equation $y = x + 1$. The general circle through O and Q has equation

$$\left(x - \frac{1}{2}\right)^2 + (y - k)^2 = \frac{1}{4} + k^2$$

or

$$x^2 - x + y^2 - 2ky = 0.$$

Solving this with $y = x + 1$ yields

$$2x^2 + (1 - 2k)x + (1 - 2k) = 0.$$

The roots of this quadratic equation are coincident if and only if $2k = 1$ or $1 - 2k = 8$. When $k = \frac{1}{2}$, we get the point of tangency $(0,1)$. When $k = -\frac{7}{2}$, we get the point of tangency $(-2,-1)$.

The first solution corresponds to a circle with chord OQ touching the line of equation $y = x + 1$ at the point $P(0,1)$. This circle has diameter PQ and the angle OPQ is equal to $45°$. This angle certainly exceeds angle OXQ, where X is any point other than P on the line in the upper half plane. The second solution corresponds to a circle touching the line of equation $y = x + 1$ below the x-axis at the point $R(-2,-1)$. The angle ORQ exceeds angle OYQ, where Y is any point other than R on the line in the lower half-plane. Now, the line $y = x + 1$ makes an angle of $45°$ with the line $y = -1$, and this angle exceeds $\angle ORQ$. Hence P is the desired point.

© Springer International Publishing Switzerland 2016
E.J. Barbeau, *University of Toronto Mathematics Competition (2001–2015)*, Problem Books in Mathematics,
DOI 10.1007/978-3-319-28106-3_5

Solution 2. Let $(t, t+1)$ be a typical point P on the line $y = x + 1$. The slope of OP is $(t+1)/t$ and of QP is $(t+1)/(t-1)$. Hence

$$\tan \angle OPQ = \frac{t+1}{2t^2 + t + 1}.$$

Let $f(t) = (t+1)/(2t^2 + t + 1)$, so that $(2t^2 + t + 1)f(t) = t + 1$. Then $(4t+1)f(t) + (2t^2 + t + 1)f'(t) = 1$, whence

$$f'(t) = -\frac{2t(t+2)}{(2t^2 + t + 1)^2}.$$

Now $f'(t) < 0$ when $t > 0$ or $t < -2$ while $f'(t) > 0$ when $-2 < t < 0$. Thus, $f(t)$ assumes its minimum value of $-1/7$ when $t = -2$ and its maximum value of 1 when $t = 0$. Hence $|f(t)| \leq 1$ with equality if and only if $t = 0$. The desired point P is $(0, 1)$.

Comment. The workings can be made a little easier by considering $\cot \angle OPQ = 2t - 1 + 2(t+1)^{-1}$. The derivative of this is $2[1 - (t+1)^2]$, and this is positive if and only if $t > 0$ and $t < -2$.

Solution 3. Let P be $(x, x+1)$ and let $\theta = \angle OPQ$. From the Law of Cosines, we have that

$$1 = x^2 + (x+1)^2 + (x-1)^2 + (x+1)^2 - 2\sqrt{(2x^2 + 2x + 1)(2x^2 + 2)} \cos \theta,$$

whence

$$\cos \theta = \frac{2x^2 + x + 1}{\sqrt{(2x^2 + 2x + 1)(2x^2 + 2)}}.$$

Let

$$f(x) = \sec^2 \theta = \frac{(2x^2 + 2x + 1)(2x^2 + 2)}{(2x^2 + x + 1)^2}$$

$$= \frac{4x^4 + 4x^3 + 6x^2 + 4x + 2}{4x^4 + 4x^3 + 5x^2 + 2x + 1}$$

$$= 1 + \frac{(x+1)^2}{(2x^2 + x + 1)^2} = 1 + g(x)^2,$$

where $g(x) = (x+1)(2x^2 + x + 1)^{-1}$. Since $g'(x) = [-2x(x+2)](2x^2 + x + 1)^{-2}$, we see that $g(x)$ has its maximum value when $x = 0$ and its minimum when $x = -2$, and vanishes when $x = -1$. Hence $g(x)^2$ and thus $\sec \theta$ assume relative maxima when $x = 0$ and $x = -2$. A quick check reveals that $x = 0$ gives the overall maximum, and so that the angle is maximized when P is located at $(0, 1)$.

2002:4. Consider the parabola of equation $y = x^2$. The normal is constructed at a variable point P and meets the parabola again in Q. Determine the location of P for which the arc length along the parabola between P and Q is minimized.

Solution. Wolog, we may assume that $u > 0$, as the arc length for u and $-u$ is the same. The tangent to the parabola at (u, u^2) has slope $2u$, and so the normal has slope $-1/2u$. The equation of the normal is

$$y - u^2 = -\frac{1}{2u}(x - u)$$

and this intersects the parabola at the point

$$\left(-u - \frac{1}{2u}, u^2 + 1 + \frac{1}{4u^2}\right).$$

The arc length is given by

$$f(u) = \int_{-u-(1/2u)}^{u} \sqrt{1 + 4x^2}\,dx = F(u) - F(-u - (1/2u)),$$

where F is a function for which $F'(x) = (1 + 4x^2)^{1/2}$. Then

$$f'(u) = F'(u) - F'(-u - (1/2u))\left(-1 + \frac{1}{2u^2}\right)$$

$$= (1 + 4u^2)^{1/2} - (1 + 4u^2 + 4 + u^{-2})^{1/2}\left(-1 + \frac{1}{2u^2}\right)$$

$$= (1 + 4u^2)^{1/2} - (4u^4 + 5u^2 + 1)^{1/2}\left(\frac{-1}{u} + \frac{1}{2u^3}\right)$$

$$= (1 + 4u^2)^{1/2}\left[1 + \frac{(u^2 + 1)^{1/2}(2u^2 - 1)}{2u^3}\right].$$

$f'(u)$ is negative when u is close to 0, and positive when u is very large. It vanishes if and only if $2u^3 = -(u^2 + 1)^{1/2}(2u^2 - 1)$. Thus $4u^6 = (u^2+1)(4u^4 - 4u^2 + 1) \Leftrightarrow 0 = -3u^2 + 1$, and we have that $f'(1/\sqrt{3}) = 0$. Hence $f(u)$ decreases on the interval $(0, 1/\sqrt{3})$ and increases on the interval $(1/\sqrt{3}, \infty)$. Hence, the arc length is minimized when P is the one of the points

$$\left(\frac{1}{\sqrt{3}}, \frac{1}{3}\right), \quad \left(-\frac{1}{\sqrt{3}}, \frac{1}{3}\right).$$

2003:3. Solve the differential equation

$$y'' = yy'.$$

Solution. The equation can be rewritten

$$\frac{dy'}{dy}y' = yy'$$

from which either $y' = 0$ and y is a constant, or $dy' = ydy$, whence

$$\frac{dy}{dx} = y' = \frac{1}{2}(y^2 + k)$$

for some constant k. (One can also get to the same place by noting that the equation can be rewritten as $2y'' = (y^2)'$ and integrating.)

In case $k = 0$, we have that $2dy/y^2 = dx$, whence $-2/y = x + c$ and $y = -2(x + c)^{-1}$ for some constant c.

In the case that $k = a^2 > 0$, we have that

$$\frac{2dy}{y^2 + a^2} = dx$$

whence $(2/a) \tan^{-1}(y/a) = x + c$ and

$$y = a \tan \frac{a(x + c)}{2}$$

for some constant c.

In the case that $k = -b^2 < 0$, we have that

$$\frac{2dy}{y^2 - b^2} = \frac{1}{b} \left[\frac{dy}{y - b} - \frac{dy}{y + b} \right] = dx$$

whence

$$\frac{1}{b} \ln \left| \frac{y - b}{y + b} \right| = x + c$$

for some constant c. Solving this for y yields that

$$y = b \left(\frac{1 + e^{b(x+c)}}{1 - e^{b(x+c)}} \right) \quad \text{and} \quad y = b \left(\frac{1 - e^{b(x+c)}}{1 + e^{b(x+c)}} \right).$$

2003:9. Prove that the integral

$$\int_0^\infty \frac{\sin^2 x}{\pi^2 - x^2} \, dx$$

exists and evaluate it.

Solution. Since $\lim_{x \to \pi} (\sin^2 x)/(\pi^2 - x^2)$ exists and equals 0 (by l'Hôpital's Rule), there is a removable singularity in the integrand at $x = \pi$. The integral on the infinite interval converges by comparison with the integral of $1/x^2$.

First, note that, for $k \geq 0$,

$$\int_{k\pi}^{(k+1)\pi} \frac{\sin^2 x}{\pi^2 - x^2} \, dx = -\frac{1}{2\pi} \left[\int_{k\pi}^{(k+1)\pi} \frac{\sin^2 x}{x - \pi} \, dx - \int_{k\pi}^{(k+1)\pi} \frac{\sin^2 x}{x + \pi} \, dx \right]$$

$$= -\frac{1}{2\pi} \int_{(k-1)\pi}^{k\pi} \frac{\sin^2 x}{x} \, dx + \frac{1}{2\pi} \int_{(k+1)\pi}^{(k+2)\pi} \frac{\sin^2 x}{x} \, dx.$$

Hence

$$\int_0^{n\pi} \frac{\sin^2 x}{\pi^2 - x^2} \, dx = -\frac{1}{2\pi} \int_{-\pi}^{\pi} \frac{\sin^2 x}{x} \, dx + \frac{1}{2\pi} \int_{(n-1)\pi}^{(n+1)\pi} \frac{\sin^2 x}{x} \, dx.$$

The first integral on the right vanishes, because the integrand is odd, and so

$$\left| \int_0^{n\pi} \frac{\sin^2 x}{\pi^2 - x^2} \, dx \right| = \frac{1}{2\pi} \int_{(n-1)\pi}^{(n+1)\pi} \frac{\sin^2 x}{x} \, dx \leq \frac{1}{2\pi} \int_{(n-1)\pi}^{(n+1)\pi} \frac{1}{x} \, dx \leq \frac{1}{(n-1)\pi},$$

with the result that $\int_0^\infty \frac{\sin^2 x}{\pi^2 - x^2} \, dx = 0$.

2004:2. Prove that

$$\int_0^1 x^x \, dx = 1 - \frac{1}{2^2} + \frac{1}{3^3} - \frac{1}{4^4} + \frac{1}{5^5} + \cdots .$$

The solution to this problem appears in Chap. 4.

2004:6. Determine

$$\left(\int_0^1 \frac{dt}{\sqrt{1 - t^4}} \right) \div \left(\int_0^1 \frac{dt}{\sqrt{1 + t^4}} \right).$$

Solution. The substitution $t^2 = \sin \theta$ leads to

$$\int_0^1 \frac{dt}{\sqrt{1 - t^4}} = \int_0^{\pi/2} \frac{\cos \theta \, d\theta}{2\sqrt{\sin \theta} \cos \theta} = \frac{1}{2} \int_0^{\pi/2} \frac{d\theta}{\sqrt{\sin \theta}}.$$

The substitution $t^2 = \tan \alpha$ followed by the substitution $\beta = 2\alpha$ leads to

$$\int_0^1 \frac{dt}{\sqrt{1 + t^4}} = \int_0^{\pi/4} \frac{\sec^2 \alpha \, d\alpha}{2\sqrt{\tan \alpha} \sec \alpha}$$

$$= \frac{1}{2} \int_0^{\pi/4} \frac{d\alpha}{\sqrt{\sin \alpha \cos \alpha}}$$

$$= \frac{1}{2} \int_0^{\pi/4} \frac{\sqrt{2} \, d\alpha}{\sqrt{\sin 2\alpha}} = \frac{1}{4} \int_0^{\pi/2} \frac{\sqrt{2} \, d\beta}{\sqrt{\sin \beta}}$$

$$= \frac{1}{2\sqrt{2}} \int_0^{\pi/2} \frac{d\beta}{\sqrt{\sin \beta}}.$$

Thus the answer is $\sqrt{2}$.

2005:1. Show that, if $-\pi/2 < \theta < \pi/2$, then

$$\int_0^\theta \log(1 + \tan \theta \tan x) \, dx = \theta \log \sec \theta.$$

Solution 1. [D. Han]

$$\int_0^\theta \log(1 + \tan\theta\tan x)\, dx = \int_0^\theta \log(1 + \tan\theta\tan(\theta - u))\, du$$

$$= \int_0^\theta \log\left[1 + \tan\theta\left(\frac{\tan\theta - \tan u}{1 + \tan\theta\tan u}\right)\right] du$$

$$= \int_0^\theta [\log(1 + \tan^2\theta) - \log(1 + \tan\theta\tan u)]\, du$$

$$= \theta\log\sec^2\theta - \int_0^\theta \log(1 + \tan\theta\tan u)\, du$$

$$= 2\theta\log\sec\theta - \int_0^\theta \log(1 + \tan\theta\tan x)\, dx.$$

The desired result follows.

Solution 2.

$$\int_0^\theta \log(1 + \tan\theta\tan x)\, dx$$

$$= \int_0^\theta [\log(\cos\theta\cos x + \sin\theta\sin x) - \log\cos\theta - \log\cos x]\, dx$$

$$= \int_0^\theta [\log\cos(\theta - x) + \log\sec\theta - \log\cos x]\, dx$$

$$= \int_0^\theta \log\cos(\theta - x)\, dx + \theta\log\sec\theta - \int_0^\theta \log\cos x\, dx$$

$$= \int_0^\theta \log\cos x dx + \theta\log\sec\theta - \int_0^\theta \log\cos x dx = \theta\log\sec\theta.$$

Solution 3. Let $F(\theta) = \int_0^\theta \log(1 + \tan\theta\tan x)\, dx$. Then, using the substitution $u = \tan x$, we find that

$$F'(\theta) = \log(1 + \tan^2\theta) + \int_0^\theta \frac{\sec^2\theta\tan x}{1 + \tan\theta\tan x}\, dx$$

$$= \log\sec^2\theta + \int_0^{\tan\theta}\left[-\frac{\tan\theta}{1 + (\tan\theta)u} + \frac{u}{u^2 + 1} + \frac{\tan\theta}{u^2 + 1}\right] du$$

$$= +2\log\sec\theta + \left[-\log(1 + (\tan\theta)u) + \frac{1}{2}\log(1 + u^2) + \tan\theta[\arctan u]\right]_0^{\tan\theta}$$

$$= 2\log\sec\theta - \log(1 + \tan^2\theta) + \frac{1}{2}\log\sec^2\theta + \theta\tan\theta$$

$$= \log\sec\theta + \theta\tan\theta.$$

Also, the derivative of $G(\theta) \equiv \theta \log \sec \theta$ is equal to $\log \sec \theta + \theta \tan \theta$. Since $F'(\theta) = G'(\theta)$ and $F(0) = G(0) = 0$, it follows that $F(\theta) = G(\theta)$ for $-\frac{\pi}{2} < \theta < \frac{\pi}{2}$, as desired.

Comment. It is interesting to observe that

$$\int_0^\theta \log(1 + \tan \theta \tan x)\, dx = \theta \cdot \frac{\log 1 + \log \sec^2 \theta}{2},$$

the length of the interval times the average of the integrand values at the endpoint.

2006:4. Two parabolas have parallel axes and intersect in two points. Prove that their common chord bisects the segments whose endpoints are the points of contact of their common tangent.

The solution to this problem appears in Chap. 8.

2007:10. Solve the following differential equation

$$2y' = 3|y|^{1/3}$$

subject to the initial conditions

$$y(-2) = -1 \quad \text{and} \quad y(3) = 1.$$

Your solution should be everywhere differentiable.

Solution. Depending on the sign of y in any region, separation of variables leads to the solution

$$y^{2/3} = x + c \quad \text{or} \quad y = (x + c)^{3/2}$$

when $y \geq 0$ and to

$$y^{2/3} = -(x + c) \quad \text{or} \quad y = -[-(x + c)]^{3/2}$$

when $y < 0$. The desired solution is

$$y(x) = \begin{cases} -[-(x + 1)]^{3/2}, & \text{if } x < -1; \\ 0, & \text{if } -1 \leq x \leq 2; \\ (x - 2)^{3/2}, & \text{if } x > 2. \end{cases}$$

[Problem **2007:10** was contributed by Victor Ivrii.]

2009:1. Determine the supremum and the infimum of

$$\frac{(x - 1)^{x-1} x^x}{(x - (1/2))^{2x-1}}$$

for $x > 1$.

Solution. Let $g(x)$ be the function in question and let

$$\begin{aligned} f(x) &= \log g(x) \\ &= (x - 1) \log(x - 1) + x \log x - (2x - 1) \log((2x - 1)/2) \\ &= (x - 1) \log(x - 1) + x \log x - (2x - 1) \log(2x - 1) + (2x - 1) \log 2. \end{aligned}$$

Then

$$f'(x) = \log(x - 1) + 1 + \log x + 1 - 2\log(2x - 1) - 2 + 2\log 2$$
$$= \log\left[\frac{4x(x - 1)}{(2x - 1)^2}\right] = \log\left[1 - \frac{1}{(2x - 1)^2}\right]$$
$$< \log 1 = 0.$$

Therefore, $f(x)$, and hence $g(x)$ is a decreasing function on $(1, \infty)$.

It is straightforward to check that $\lim_{x\downarrow 1} f(x) = \log 2$, so that $\lim_{x\downarrow 1} g(x) = 2$.

To check behaviour for large values of x, let

$$h(u) = g(u + (1/2))$$
$$= \frac{(u - (1/2))^{u - (1/2)}(u + (1/2))^{u + (1/2)}}{u^{2u}}$$
$$= \left(1 - \frac{1}{2u}\right)^u \left(1 + \frac{1}{2u}\right)^u \left(1 - \frac{1}{2u}\right)^{-1/2} \left(1 + \frac{1}{2u}\right)^{1/2}$$
$$= \left(1 - \frac{1}{4u^2}\right)^u \left(\frac{1 + (1/2u)}{1 - (1/2u)}\right)^{1/2}.$$

When $v = 1/(2u)$, the logarithm of the first factor is

$$\frac{\log(1 - v^2)}{2v},$$

and an application of l'Hôpital's Rule yields that its limit is 0 as $v \to 0$. It follows that

$$\lim_{u \to \infty} h(u) = 1.$$

Therefore, the desired supremum is 2 and infimum is 1.

2012:8. Determine the area of the set of points (x, y) in the plane that satisfy the two inequalities:

$$x^2 + y^2 \leq 2$$
$$x^4 + x^3 y^3 \leq xy + y^4.$$

The solution to this problem appears in Chap. 8.

2015:5. Let $f(x)$ be a real polynomial of degree 4 whose graph has two real inflection points. There are three regions bounded by the graph and the line passing through these inflection points. Prove that two of these regions have equal area and that the area of the third region is equal to the sum of the other two areas.

Solution. By scaling and translating, we may assume that the two inflection points are located at $x = -1$ and $x = 1$ and that $f(x)$ is monic. Since $f''(x)$ is a multiple of $(x+1)(x-1) = x^2 - 1$, we have that

$$f(x) = (x^4 - 6x^2 + 5) + (bx + c)$$

where $2b = f(1) - f(-1)$ and $2c = f(1) + f(-1)$. The line passing through the inflection points $(-1, f(-1))$ and $(1, f(1))$ is $y = g(x)$ with $g(x) = bx + c$. Since $f(x) - g(x) = (x^2 - 5)(x^2 - 1)$, the curves with equations $y = f(x)$ and $y = g(x)$ intersect when $x = \pm\sqrt{5}, \pm1$. The three areas in question are given by

$$\left| \int_{-\sqrt{5}}^{-1} (x^4 - 6x^2 + 5)\, dx \right| = \left| \int_1^{\sqrt{5}} (x^4 - 6x^2 + 5)\, dx \right|$$

$$= \left| \left[x^5/5 - 2x^3 + 5x \right]_1^{\sqrt{5}} \right| = \frac{16}{5}$$

and

$$\int_{-1}^1 (x^4 - 6x^2 + 5)\, dx = \left[x^5/5 - 2x^3 + 5x \right]_{-1}^1 = 32/5.$$

The result follows.

[Problem **2015:5** is #E817 from the *American Mathematics Monthly* 55:5 (May, 1948), 317; 56:2 (February, 1949), 106–108.]

2015:7. Determine

$$\int_0^2 \frac{e^x\, dx}{e^{1-x} + e^{x-1}}.$$

Solution 1.

$$\int_0^2 \frac{e^x\, dx}{e^{1-x} + e^{x-1}} = e \int_{-1}^1 \frac{e^u\, du}{e^{-u} + e^u} = e \int_{-1}^1 \frac{e^{2u}\, du}{e^{2u} + 1}$$

$$= \frac{e}{2} \left[\log(e^{2u} + 1) \right]_{-1}^1 = \frac{e}{2} \left[\log\left(\frac{e^2 + 1}{e^{-2} + 1} \right) \right]$$

$$= \frac{e}{2} \log e^2 = e.$$

Solution 2. Setting $u = e^x$, we find that the integral is equal to

$$\int_1^{e^2} \frac{du}{(e/u) + (u/e)} = e \int_1^{e^2} \frac{u\, du}{e^2 + u^2} = \frac{e}{2} \left[\log(e^2 + u^2) \right]_1^{e^2}$$

$$= \frac{e}{2} \log\left(\frac{e^2 + e^4}{e^2 + 1} \right) = \frac{e}{2} \log e^2 = e.$$

Solution 3. We first establish a general result: *Suppose that f is continuous on the interval $[a, b]$. Then*

$$\int_a^b \frac{f(x - a)\, dx}{f(x - a) + f(b - x)} = \frac{1}{2}(b - a).$$

Making the substitution $x = a + b - u$, we see that the given integral is equal to

$$\int_a^b \frac{f(b-u)\,du}{f(u-a)+f(b-u)}.$$

Adding the two integrals together yields $\int_a^b dx = b - a$, from which the result is found.

Apply this result to $f(x) = e^x$, $a = 0$ and $b = 2$, to obtain

$$\int_0^2 \frac{e^x\,dx}{e^{1-x}+e^{x-1}} = e\int_0^2 \frac{e^x\,dx}{e^{2-x}+e^x} = e.$$

CHAPTER 6

Other Topics in Analysis

2001:7. Suppose that $x \geq 1$ and that $x = \lfloor x \rfloor + \{x\}$, where $\lfloor x \rfloor$ is the greatest integer not exceeding x and the fractional part $\{x\}$ satisfies $0 \leq \{x\} < 1$. Define

$$f(x) = \frac{\sqrt{\lfloor x \rfloor} + \sqrt{\{x\}}}{\sqrt{x}}.$$

   (a) Determine the supremum, i.e., the least upper bound, of the values of $f(x)$ for $1 \leq x$.
   (b) Let $x_0 \geq 1$ be given, and for $n \geq 1$, define $x_n = f(x_{n-1})$. Prove that $\lim_{n \to \infty} x_n$ exists.

Solution. (a) Let $x = y + z$, where $y = \lfloor x \rfloor$ and $z = \{x\}$. Then

$$f(x)^2 = 1 + \frac{2\sqrt{yz}}{y + z},$$

which is less than 2 because $y \neq z$ and $\sqrt{yz} < \frac{1}{2}(y + z)$ by the arithmetic-geometric means inequality. Hence $0 \leq f(x) < \sqrt{2}$ for each value of x. Taking $y = 1$, we find that

$$\lim_{x \uparrow 2} f(x)^2 = \lim_{z \uparrow 1} \left(1 + \frac{2\sqrt{z}}{1 + z}\right) = 2,$$

whence $\sup\{f(x) : x \geq 1\} = \sqrt{2}$.
   (b) In determining the fate of $\{x_n\}$, note that after the first entry, the sequence lies in the interval $[1, 2)$. So, wolog, we may assume that $1 \leq x_0 < 2$. If $x_0 = 1$, then each $x_n = 1$ and the limit is 1. For the rest, note that $f(x)$ simplifies to $(1 + \sqrt{x - 1})/\sqrt{x}$ on $(1, 2)$. The key point now is to observe that there is exactly one value v between 1 and 2 for which $f(v) = v$.

   Let $x = 1 + u$ with $u > 0$. Then it can be checked that $f(x) = x$ if and only if $1 + 2\sqrt{u} + u = 1 + 3u + 3u^2 + u^3$ or $u^5 + 6u^4 + 13u^3 + 12u^2 + 4u - 4 = 0$. Since the left side is strictly increasing in u, takes the value -4 when $u = 0$

© Springer International Publishing Switzerland 2016
E.J. Barbeau, *University of Toronto Mathematics Competition (2001–2015)*, Problem Books in Mathematics,
DOI 10.1007/978-3-319-28106-3_6

and the value 32 when $u = 1$, the equation is satisfied for exactly one value of u in $(0, 1)$; now let $v = 1 + u$ for this value of u. The value of v turns out to be approximately 1.375.

Since

$$f'(x) = \frac{1}{2}x^{-3/2}(x-1)^{-1/2}[(2x-1) - (x-1)^{1/2}] > 0,$$

the function $f(x)$ is strictly increasing on $(1, 2)$. Observing that $f(5/4) = 3/\sqrt{5} > 5/4$ and $f(25/16) = 7/5 < 25/16$, we conclude that, when $1 < x < v$, $x < f(x) < f(v) = v$, and when $v < x < 2$, then $v = f(v) < f(x) < x$. Thus, when $1 < x_0 < v$, the iterates $\{x_n\}$ constitute a bounded increasing sequence whose limit is a fixed point of f, and when $v < x_0 < 2$, the iterates $\{x_n\}$ constitute a bounded decreasing sequence whose limit is a fixed point of f. In either case, $\lim_{n\to\infty} x_n = v$.

2002:5. Let n be a positive integer. Suppose that f is a function defined and continuous on $[0, 1]$ that is differentiable on $(0, 1)$ and satisfies $f(0) = 0$ and $f(1) = 1$. Prove that, there exist n [distinct] numbers x_i $(1 \le i \le n)$ in $(0, 1)$ for which

$$\sum_{i=1}^{n} \frac{1}{f'(x_i)} = n.$$

Solution. Since $f(x)$ is continuous on $[0, 1]$, it assumes every value between 0 and 1 inclusive. Select points $0 = u_0 < u_1 < u_2 < \cdots < u_{n-1} < u_n = 1$ in $[0, 1]$ for which $f(u_i) = i/n$ for $0 \le i \le n$. Then, by the Mean Value Theorem, for each $i = 1, 2, \ldots, n$, there exists $x_i \in (u_{i-1}, u_i)$ for which

$$\frac{1}{n(u_i - u_{i-1})} = \frac{f(u_i) - f(u_{i-1})}{u_i - u_{i-1}} = f'(x_i).$$

Therefore,

$$\sum_{i=1}^{n} \frac{1}{f'(x_i)} = n \sum_{i=1}^{n} (u_i - u_{i-1}) = n.$$

2003:5. For $x > 0$, $y > 0$, let $g(x, y)$ denote the minimum of the three quantities, x, $y + 1/x$ and $1/y$. Determine the maximum value of $g(x, y)$ and where this maximum is assumed.

Solution 1. Consider Fig. 6.1 showing the curves of equations $x = y + 1/x$, $x = 1/y$ and $y + 1/x = 1/y$, all three of which contain the point $P \sim (\sqrt{2}, 1/\sqrt{2})$, and the regions in which $g(x, y)$ is each one of the three given functions. In region **A**, $g(x, y) = x$. In region **B**, $g(x, y) = 1/y$. In region **C**, $g(x, y) = y + 1/x$.

When $g(x, y) = x$, the maximum value of $g(x, y)$ is equal to $\sqrt{2}$, assumed at the point P. When $g(x, y) = 1/y$, the maximum value of $g(x, y)$ is equal to $\sqrt{2}$, also assumed at P. Suppose (x, y) is such that $g(x, y) = y + 1/x$. Note that the curve $y = \sqrt{2} - (1/x)$ passes through the point P, where it intersects each of the curves $y + (1/x) = (1/y)$ and $y + (1/x) = x$. It intersects neither of these curves at any other point, and so lies vertically above the region where $g(x, y) = y + (1/x)$. In this region, $y \leq \sqrt{2} - (1/x)$ and so $g(x, y) \leq \sqrt{2}$, with equality only at the point P. Hence the required maximum is $\sqrt{2}$, and it is assumed at $(x, y) = (\sqrt{2}, 1/\sqrt{2})$.

Solution 2. [R. Appel] If $x \leq 1$ and $y \leq 1$, then $g(x, y) \leq x \leq 1$. If $y \geq 1$, then $g(x, y) \leq 1/y \leq 1$. It remains to examine the case $x > 1$ and $y < 1$, so that $y + (1/x) < 2$. Suppose that min $(x, 1/y) = a$ and max $(x, 1/y) = b$. Then max $(1/x, y) = 1/a$ and min $(1/x, y) = 1/b$, so that

$$y + \frac{1}{x} = \frac{1}{a} + \frac{1}{b} = \frac{a+b}{ab}.$$

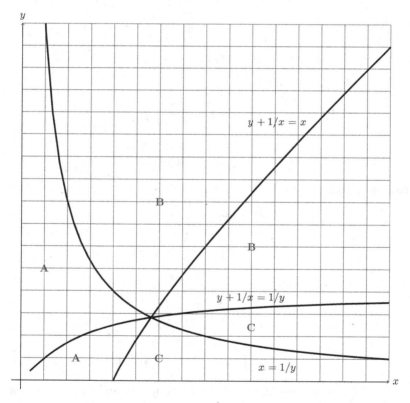

Fig. 6.1. In region **A**, $g(x, y) = x$. In region **B**, $g(x, y) = 1/y$. In region **C**, $g(x, y) = y + 1/x$

Hence $g(x,y) = \min{(a, (a+b)/(ab))}$. Either $a^2 \leq 2$ or $a^2 \geq 2$. But in the latter case,

$$\frac{a+b}{ab} \leq \frac{2b}{\sqrt{2}b} = \sqrt{2}.$$

In either case, $g(x,y) \leq \sqrt{2}$. This maximum value is attained when $(x,y) = (\sqrt{2}, 1/\sqrt{2})$.

Solution 3. [D. Varodayan] By the continuity of the functions, each of the regions $\{(x,y) : 0 < x < y + (1/x), xy < 1\}$, $\{(x,y) : 0 < x, y + (1/x) < x, y + (1/x) < (1/y)\}$, and $\{(x,y) : 0 < (1/y) < x, (1/y) < y + (1/x)\}$ is an open subset of the plane; using partial derivatives, we see that none of the three functions being minimized have any critical values there. It follows that any extreme values of $g(x,y)$ must occur on one of the curves defined by the equations

(6.1) $$x = y + (1/x)$$

(6.2) $$x = 1/y$$

(6.3) $$y + (1/x) = (1/y)$$

On the curve (6.1), $x \geq 1$ and

$$g(x,y) = \min\left(x, \frac{x}{x^2-1}\right)$$

$$= \begin{cases} x, & \text{if } x \leq \sqrt{2}; \\ \frac{x}{x^2-1}, & \text{if } x \geq \sqrt{2}. \end{cases}$$

On the curve (6.2),

$$g(x,y) = \min(x, 2/x)$$

$$= \begin{cases} x, & \text{if } x \leq \sqrt{2}; \\ 2/x, & \text{if } x \geq \sqrt{2}. \end{cases}$$

On the curve (6.3), $0 < y < 1$ and

$$g(x,y) = \min\left(\frac{y}{1-y^2}, \frac{1}{y}\right)$$

$$= \begin{cases} \frac{y}{1-y^2}, & \text{if } 0 < y < \frac{1}{\sqrt{2}}; \\ 1/y, & \text{if } \frac{1}{\sqrt{2}} \leq y \leq 1. \end{cases}$$

On each of these curves, $g(x,y)$ reaches its maximum value of $\sqrt{2}$ when $(x,y) = (\sqrt{2}, 1/\sqrt{2})$.

2003:7. Suppose that the polynomial $f(x)$ of degree $n \geq 1$ has all real roots and that $\lambda > 0$. Prove that the set $\{x \in \mathbb{R} : |f(x)| \leq \lambda|f'(x)|\}$ is a finite union of closed intervals whose total length is equal to $2n\lambda$.

Solution. Wolog, we may assume that the leading coefficient is 1. Let $f(x) = \prod_{i=1}^{k}(x - r_i)^{m_i}$, where $n = \sum_{i=1}^{k} m_i$. Then

$$\frac{f'(x)}{f(x)} = \sum_{i=1}^{k} \frac{m_i}{x - r_i}.$$

Note that the derivative of this function, $-\sum_{i=1}^{k} m_i(x - r_i)^{-2}$ is less than 0, so that it decreases on each interval upon which it is defined. By considering the graph of $f'(x)/f(x)$, we see that $f'(x)/f(x) \geq 1/\lambda$ on finitely many intervals of the form $(r_i, s_i]$, where $r_i < s_i$ and the r_i and s_j interlace, and $f'(x)/f(x) \leq -1/\lambda$ on finitely many intervals of the form $[t_i, r_i)$, where $t_i < r_i$ and the t_i and r_j interlace. For each i, we have $t_i < r_i < s_i < t_{i+1}$.

The equation $f'(x)/f(x) = 1/\lambda$ can be rewritten as

$$0 = (x - r_1)(x - r_2)\cdots(x - r_k) - \lambda \sum_{i=1}^{k} m_i(x - r_1)\cdots\widehat{(x - r_i)}\cdots(x - r_k)$$

$$= x^k - \left(\sum_{i=1}^{k} r_i - \lambda \sum_{i=1}^{k} m_i\right) x^{k-1} + \cdots.$$

The sum of the solutions of this equation is

$$s_1 + s_2 + \cdots + s_k = r_1 + \cdots + r_k + \lambda n,$$

so that $\sum_{i=1}^{m}(s_i - r_i) = \lambda n$. This is the sum of the lengths of the intervals $(r_i, s_i]$ on which $f'(x)/f(x) \geq 1/\lambda$. Similarly, we can show that $f'(x)/f(x) \leq -1/\lambda$ on a finite collection of intervals of total length λn. The set on which the inequality of the problem holds is equal to the union of all of these half-open intervals and the set $\{r_1, r_2, \ldots, r_k\}$. The result follows.

2005:2. Suppose that f is continuously differentiable on $[0, 1]$ and that $\int_0^1 f(x)dx = 0$. Prove that

$$2\int_0^1 f(x)^2\,dx \leq \int_0^1 |f'(x)|\,dx \cdot \int_0^1 |f(x)|\,dx.$$

Solution 1.

$$2\int_0^1 f(x)^2\,dx = \int_0^1 f(x)[2f(x) - f(0) - f(1)]\,dx$$

$$= \int_0^1 f(x)[(f(x) - f(0)) - (f(1) - f(x))]\,dx$$

$$= \int_0^1 f(x)\left[\int_0^x f'(t)dt - \int_x^1 f'(t)\,dt\right]dx$$

$$\leq \int_0^1 |f(x)|\left|\int_0^x f'(t)\,dt - \int_x^1 f'(t)\,dt\right|dx$$

$$\leq \int_0^1 |f(x)| \left(\int_0^x |f'(t)|\, dt + \int_x^1 |f'(t)|\, dt \right) dx$$

$$= \int_0^1 |f(x)| \left(\int_0^1 |f'(t)|\, dt \right) dx$$

$$= \int_0^1 |f'(x)|\, dx \cdot \int_0^1 |f(x)|\, dx.$$

Solution 2. Let $F(x) = \int_0^x f(t)\, dt$. Then, integrating by parts, we find that

$$\int_0^1 f^2(x)\, dx = [f(x)F(x)]_0^1 - \int_0^1 f'(x)F(x)\, dx = -\int_0^1 f'(x)F(x)\, dx.$$

Hence

$$\int_0^1 f^2(x)\, dx \leq \int_0^1 |f'(x)||F(x)|\, dx \leq \left(\int_0^1 |f'(x)|\, dx \right) \cdot \sup\{|F(x)| : 0 \leq x \leq 1\}.$$

Now

$$2|F(x)| = |F(x) - F(0)| + |F(1) - F(x)| = \left| \int_0^x f(t)\, dt \right| + \left| \int_x^1 f(t)\, dt \right|$$

$$\leq \int_0^1 \left| f(t) \right| dt,$$

from which the result follows.

2005:5. Let $f(x)$ be a polynomial with real coefficients, evenly many of which are nonzero, which is *palindromic*. This means that the coefficients read the same in either direction, i.e. $a_k = a_{n-k}$ if $f(x) = \sum_{k=0}^n a_k x^k$, or alternatively, $f(x) = x^n f(1/x)$, where n is the degree of the polynomial. Prove that $f(x)$ has at least one root of absolute value 1.

Solution. If $n = \deg f(x)$ is odd, then k and $n - k$ have opposite parity and

$$f(-1) = \sum \{a_k((-1)^k + (-1)^{n-k}) : 0 \leq k < n/2\} = 0.$$

Suppose that $n = 2m$. Then $f(x)$ has $2m + 1$ coefficients: m pairs (a_k, a_{n-k}) of equal coefficients ($0 \leq k \leq m - 1$) and a_m. Since evenly many coefficients are nonzero, we must have that $a_m = 0$. Hence

$$f(x) = \sum_{k=0}^{m-1} a_k(x^k + x^{2m-k})$$

$$= x^m \sum_{k=0}^{m-1} a_k(x^{-(m-k)} + x^{m-k})$$

$$= x^m \sum_{j=1}^{m} a_{m-j}(x^{-j} + x^j).$$

Then $f(e^{i\theta}) = 2e^{mi\theta} \sum_{j=1}^{m} a_{m-j} \cos j\theta$. Let $P(\theta) = \sum_{j=1}^{m} a_{m-j} \cos j\theta$. Then $e^{i\theta}$ is a root of f if and only if $P(\theta) = 0$.

Now $\int_0^{2\pi} P(\theta)d\theta = 0$ and $P(\theta)$ is not constant (otherwise f would be identically 0), so $P(\theta)$ is a continuous real function that assumes both positive and negative values. Hence, by the Intermediate Value Theorem, P must vanish somewhere on the interval $[0, 2\pi]$, and the result follows.

2005:9. Let S be the set of all real-valued functions that are defined, positive and twice continuously differentiable on a neighbourhood of 0. Suppose that a and b are real parameters with $ab \neq 0$, $b < 0$. Define operators from S to $\mathbb{R}$ as follows:

$$A(f) = f(0) + af'(0) + bf''(0);$$

$$G(f) = \exp A(\log f).$$

(a) Prove that $A(f) \leq G(f)$ for $f \in S$;

(b) Prove that $G(f + g) \leq G(f) + G(g)$ for $f, g \in S$;

(c) Suppose that H is the set of functions in S for which $G(f) \leq f(0)$. Give examples of nonconstant functions, one in H and one not in H. Prove that, if $\lambda > 0$ and $f, g \in H$, then λf, $f + g$ and fg all belong to H.

Solution. (a) The result if clear if $A(f) \leq 0$. Suppose that $A(f) > 0$. We have that $(\log f)' = f'/f$ and $(\log f)'' = (f''/f) - (f'/f)^2$, so that

$$A(\log f) = \log f(0) + \frac{af'(0)}{f(0)} + \frac{bf''(0)}{f(0)} - b\left(\frac{f'(0)}{f(0)}\right)^2.$$

We have that

$$\log A(f) = \log f(0) + \log\left(1 + \frac{af'(0)}{f(0)} + \frac{bf''(0)}{f(0)}\right)$$

$$\leq \log f(0) + \frac{af'(0)}{f(0)} + \frac{bf''(0)}{f(0)}$$

$$= A(\log f) + b\left(\frac{f'(0)}{f(0)}\right)^2 \leq A(\log f).$$

(b) Note that, for any functions $u, v \in S$,

$$G(uv) = \exp A(\log u + \log v) = \exp[A(\log u) + A(\log v)]$$

$$= \exp A(\log u) \cdot \exp A(\log v) = G(u)G(v).$$

Similarly $G(u/v) = G(u)/G(v)$. From (a), we have the inequalities

$$A\left(\frac{f}{f+g}\right) \leq G\left(\frac{f}{f+g}\right) \quad \text{and} \quad A\left(\frac{g}{f+g}\right) \leq G\left(\frac{g}{f+g}\right).$$

Adding these two inequalities yields

$$1 \leq \frac{G(f) + G(g)}{G(f+g)}$$

from which the desired result follows.
(c) Suppose that $\lambda > 0$ and $f, g \in H$. Then

$$G(\lambda f) = G(\lambda)G(f) = \lambda G(f) \leq \lambda f(0);$$

$$G(fg) = G(f)G(g) \leq f(0)g(0) = (fg)(0);$$

$$G(f + g) \leq G(f) + G(g) \leq f(0) + g(0) = (f + g)(0).$$

Suppose that $a \neq 0$, and let $f(x) = e^{kx}$. Then $G(f) = e^{ka}$. If $ka > 0$, then $f \notin H$; if $ka < 0$, then $f \in H$. Suppose that $a = 0$, and let $f(x) = e^{kx^2}$. Then $G(f) = e^{2kb}$. If $k < 0$, then $f \notin H$; if $k > 0$, then $f \in H$.

2006:3. Let $p(x)$ be a polynomial of positive degree n with n distinct real roots $a_1 < a_2 < \cdots < a_n$. Let b be a real number for which $2b < a_1 + a_2$. Prove that

$$2^{n-1}|p(b)| \geq |p'(a_1)(b - a_1)|.$$

The solution to this problem appears in Chap. 2.

2006:6. Suppose that k is a positive integer and that

$$f(t) = a_1 e^{\lambda_1 t} + a_2 e^{\lambda_2 t} + \cdots + a_k e^{\lambda_k t}$$

where $a_1, \ldots, a_k, \lambda_1, \ldots, \lambda_k$ are real numbers with $\lambda_1 < \lambda_2 < \cdots < \lambda_k$. Prove that $f(t) = 0$ has finitely many real solutions. What is the maximum number of solutions possible, as a function of k?

Solution. We prove by induction that the equation can have at most $k - 1$ solutions. It holds for $k = 1$. Suppose that it holds up to k, and that $a_1 e^{\lambda_1 t} + \cdots + a_{k+1} e^{\lambda_{k+1} t}$ vanishes for m distinct values of t. Then

$$g(t) \equiv a_1 e^{(\lambda_1 - \lambda_{k+1})t} + a_2 e^{(\lambda_2 - \lambda_{k+1})t} + \cdots + a_k e^{(\lambda_k - \lambda_{k+1})t} + a_{k+1}$$

has the same roots, so that, by Rolle's theorem,

$$g'(t) \equiv a_1(\lambda_1 - \lambda_{k+1})e^{(\lambda_1 - \lambda_{k+1})t} + a_2(\lambda_2 - \lambda_{k+1})e^{(\lambda_2 - \lambda_{k+1})t}$$
$$+ \cdots + a_k(\lambda_k - \lambda_{k+1})e^{(\lambda_k - \lambda_{k+1})t}$$

has $m - 1$ distinct roots. Hence $m - 1 \leq k - 1$ and $m \leq k$.

For $k \geq 2$, let $p_k(x) = (x - 1)(x - 2)(x - 3) \cdots (x - \overline{k - 1}) \equiv a_1 + a_2 x + \cdots + a_k x^{k-1}$. Then $f_k(t) = e^t p_k(e^t) = a_1 e^t + \cdots + a_k e^{kt}$ has $k - 1$ roots, namely $t_i = \log i$ $(1 \leq i \leq k - 1)$. Hence $k - 1$ roots are possible.

2006:8. Let $f(x)$ be a real function defined and twice differentiable on an open interval containing $[-1, 1]$. Suppose that $0 < \alpha \leq \gamma$ and that $|f(x)| \leq \alpha$ and $|f''(x)| \leq \gamma$ for $-1 \leq x \leq 1$. Prove that

$$|f'(x)| \leq 2\sqrt{\alpha\gamma}$$

for $-1 \leq x \leq 1$. (Part marks are possible for the weaker inequality $|f'(x)| \leq \alpha + \gamma$.)

Solution. Let $|f'(x)|$ attain its maximum value on $[-1, 1]$ at $x = u$. Then, for $-1 \leq a \leq u \leq b \leq 1$, we have, by Taylor's theorem, for some v, w in $[a, b]$,

$$f(a) - f(u) = (a - u)f'(u) + \frac{1}{2}(a - u)^2 f''(v),$$

$$f(b) - f(u) = (b - u)f'(u) + \frac{1}{2}(b - u)^2 f''(w).$$

Therefore

$$(b - a)f'(u) = f(b) - f(a) + \frac{1}{2}[(a - u)^2 f''(v) - (b - u)^2 f''(w)]$$

so that

$$(b - a)|f'(u)| \leq |f(a)| + |f(b)| + \frac{1}{2}[(b - u)^2 |f''(w)| + (a - u)^2 |f''(v)|]$$

$$\leq 2\alpha + \frac{\gamma}{2}[(b - u)^2 + (a - u)^2] \leq 2\alpha + \frac{\gamma}{2}(b - a)^2,$$

since

$$(b - a)^2 - (b - u)^2 - (a - u)^2 = 2(b - u)(u - a) \geq 0.$$

Select a and b so that $b - a = 2\sqrt{\alpha/\gamma}$. Then

$$|f'(u)| \leq \left(\sqrt{\frac{\gamma}{\alpha}}\right)(\alpha) + \gamma\sqrt{\frac{\alpha}{\gamma}} = 2\sqrt{\alpha\gamma}.$$

Comment. If $a = -1$ and $b = 1$, then we get in the above that $2|f'(u)| \leq 2\alpha + 2\gamma$.

2007:4. Suppose that $f(x)$ is a continuous real-valued function defined on the interval $[0, 1]$ that is twice differentiable on $(0, 1)$ and satisfies (i) $f(0) = 0$ and (ii) $f''(x) > 0$ for $0 < x < 1$.

  (a) Prove that there exists a number a for which $0 < a < 1$ and $f'(a) < f(1)$;
  (b) Prove that there exists a unique number b for which $a < b < 1$ and $f'(a) = f(b)/b$.

Solution. (a) By the Mean Value Theorem, there exists $c \in (0, 1)$ for which $f'(c) = f(1)$. Since $f'(x)$ is increasing, when $0 < a < c$, $f'(a) < 1$.

  (b) For $a \in (0, c)$, let $g(x) = f(x) - xf'(a)$. Then $g(0) = 0$ and $g'(x) = f'(x) - f'(a)$. For $0 < u < a < v < 1$, $g'(u) < 0 < g'(v)$ (since $f'(x)$ increases). Therefore $g(a) < 0$ and $g(1) > 0$. Since $g'(x) > 0$ for $a < x < 1$, there is a unique number b for which $g(b) = 0$, and the result follows.

2007:5. For $x \leq 1$ and $x \neq 0$, let

$$f(x) = \frac{-8[1 - (1-x)^{1/2}]^3}{x^2}.$$

(a) Prove that $\lim_{x \to 0} f(x)$ exists. Take this as the value of $f(0)$.

(b) Determine the smallest closed interval that contains the set of all values assumed by $f(x)$ on its domain.

(c) Prove that $f(f(f(x))) = f(x)$ for all $x \leq 1$.

Solution. Suppose that $x = 1 - t^2$ for $t \geq 0$ and let $g(t) = f(1 - t^2)$. Then

$$g(t) = \frac{-8(1-t)^3}{(1-t)^2(1+t)^2} = \frac{8(t-1)}{(t+1)^2} = \frac{8}{(t+1)} - \frac{16}{(t+1)^2}.$$

(a) Since $g(1) = 0$, its follows that $\lim_{x \to 0} f(x) = 0$.

(b) Since

$$0 \leq \left(1 - \frac{4}{t+1}\right)^2 = 1 - \left(\frac{8}{t+1} - \frac{16}{(t+1)^2}\right),$$

$g(t)$ assumes its maximum value of 1 at $t = 3$. Indeed,

$$g'(t) = -8(t+1)^{-2} + 32(t+1)^{-3} = 8(t+1)^{-3}(3-t),$$

so that $g(t)$ increases from -8 at $t = 0$ to its maximum at $t = 3$ and then decreases to the limit 0 as t tends to infinity. Therefore the smallest closed interval containing the image of f is $[-8, 1]$. Observe that this interval gets mapped onto itself one-to-one, with $f(-8) = 1$ and $f(1) = -8$.

(c) Let $x = 8(t-1)(t+1)^{-2} = f(1 - t^2)$. Then

$$1 - x = (t-3)^2(t+1)^{-2},$$

so that

$$(1-x)^{1/2} = \begin{cases} \frac{3-t}{1+t}, & \text{if } 0 \leq t \leq 3; \\ \frac{t-3}{1+t}, & \text{if } t > 3. \end{cases}$$

and

$$1 - (1-x)^{1/2} = \begin{cases} \frac{2(t-1)}{t+1}, & \text{if } 0 \leq t \leq 3; \\ \frac{4}{t+1}, & \text{if } t > 3. \end{cases}$$

Thus

$$f(f(1 - t^2)) = f\left[\frac{8(t-1)}{(t+1)^2}\right] = \begin{cases} 1 - t^2, & \text{if } 0 \leq t \leq 3; \\ \frac{-8(t+1)}{(t-1)^2}, & \text{if } t > 3. \end{cases}$$

Thus, $f(f(x)) = x$ when $-8 \leq x \leq 1$, i.e. when $x = 1 - t^2$ for $0 \leq t \leq 3$.

Since $-8 \leq f(x) \leq 1$ for all $x \leq 1$, it follows that $f(f(f(x))) = f(x)$ for all $x \leq 1$.

Comment. It is of slight interest to note that $f(-3) = f(-24) = 8/9$.

2008:2.

(a) Determine a real-valued function g defined on the real numbers that is decreasing and for which $g(g(x)) = 2x + 2$.

(b) Prove that there is no real-valued function f defined on the real numbers that is decreasing and for which $f(f(x)) = x + 1$.

(a) *Solution.* Let $r > 0$, $r \neq 1$. We determine a decreasing composite square root of $r^2x + r^2$. Let $g(x) = -rx + b$. Then $g(g(x)) = r^2x + b(1 - r)$. When $b = r^2(1-r)^{-1}$, g is a decreasing function for which $g(g(x)) = r^2x + r^2$. The particular case in the problem can be thus dealt with; an answer is

$$g(x) = -\sqrt{2}x - 2(\sqrt{2} + 1) = -\sqrt{2}(x + 2) - 2.$$

(b) *Solution 1.* [C. Ochanine] Suppose that $f(f(x)) = x + 1$ ($f(x) = x + \frac{1}{2}$ defines such a function). Then

$$f(x + 1) = f(f(f(x))) = f(x) + 1$$

for all x, whence $f(x + 1) > f(x)$. Therefore, f cannot be decreasing.

Solution 2. If there is such a function f, then it must be one-one and onto $\mathbb{R}$. It must be continuous, since the only discontinuities a monotone function can have are jump discontinuities. Observe that $u <> f(u)$ if and only if $v = f(u) >< f(f(u)) = f(v)$. It follows that $x - f(x)$ can take both positive and negative values. By the Intermediate Value Theorem, there is a number z for which $z = f(z)$. But then $z = f(f(z)) = z + 1$, a contradiction. Therefore there is no such function f.

Solution 3. Suppose, if possible, there is a decreasing continuous function f for which $f(f(x)) = x + 1$. Since $x + 1$ has no fixpoint, neither does the function $f(x)$. Therefore $f(x) - x$ never vanishes. As f is continuous on $\mathbb{R}$, either $f(x) > x$ for all real x or $f(x) < x$ for all real x.

Suppose the former. Let M exceed the maximum of 0 and $f(0)$. Then $f(0) > f(M) > M > f(0)$, a contradiction. Therefore, $f(x) > x$ for all x cannot occur. Similarly, $f(x) < x$ for all x cannot occur. Therefore, no such function f exists and $x + 1$ has no continuous decreasing composite square root.

2010:7. Suppose that f is a continuous real-valued function defined on the closed interval $[0, 1]$ and that

$$\left(\int_0^1 xf(x) \, dx \right)^2 = \left(\int_0^1 f(x) \, dx \right) \left(\int_0^1 x^2 f(x) \, dx \right).$$

Prove that there is a point $c \in (0, 1)$ for which $f(c) = 0$.

Solution 1. Suppose, if possible, that f never vanishes on the interval, then it must be everywhere positive or negative. By replacing f by $-f$ is necessary, wolog we can assume that $f(x) > 0$ on $[0,1]$. Let $f(x) = [g(x)]^2$ for some positive function g. Then the equation becomes

$$\left(\int_0^1 xg^2(x)\,dx\right)^2 = \left(\int_0^1 g^2(x)\,dx\right)\left(\int_0^1 (xg(x))^2\,dx\right).$$

This is the equality situation in the Cauchy-Schwarz Inequality, whence $xg(x)$ must be a constant multiple of $g(x)$. But this is not the case. Therefore, by a contradiction argument, the result follows.

Solution 2. The condition and conclusion is satisfied by the zero function. Suppose, henceforth, that f is not identically zero and let $\int_0^1 f(x)\,dx = u$, $\int_0^1 xf(x)\,dx = uv$. If $u = 0$, then f must assume both positive and negative values in $(0,1)$, and so, by the Intermediate Value Theorem, must vanish. Assume $u \neq 0$. Then the condition implies that $\int_0^1 x^2 f(x)\,dx = uv^2$. Then

$$\int_0^1 (x-v)^2 f(x)\,dx = \int_0^1 x^2 f(x)\,dx - 2v\int_0^1 xf(x)\,dx + v^2\int_0^1 f(x)\,dx$$
$$= uv^2 - 2uv^2 + uv^2 = 0,$$

whence $(x-v)^2 f(x)$, and also $f(x)$ must assume both positive and negative values on $(0,1)$. The result follows.

2010:9. Let f be a real-valued function defined on $\mathbb{R}$ with a continuous third derivative, let $S_0 = \{x : f(x) = 0\}$, and, for $k = 1,2,3$, $S_k = \{x : f^{(k)}(x) = 0\}$, where $f^{(k)}$ denotes the kth derivative of f. Suppose also that $\mathbb{R} = S_0 \cup S_1 \cup S_2 \cup S_3$. Must f be a polynomial of degree not exceeding 2?

Solution. Observe that, because f and its derivatives are continuous, each set S_k is closed. Note also that, if U is an open subset upon which f or either of its first two derivatives vanish, then all derivatives of higher order (in particular, $f^{(3)}$ must also vanish on U).

First, we show that S_3 is dense in $\mathbb{R}$. Otherwise, there is an open interval J upon which $f^{(3)}$ never vanishes, so that $J \subseteq S_0 \cup S_1 \cup S_2$. The set $J \setminus S_2$ must be a nonvoid open set, and so contain an open interval J_1 upon which f'' never vanishes. Similarly, there is a nonvoid open interval $J_2 \subseteq J_1$ upon which f' never vanishes and a nonvoid open interval $J_3 \subseteq J_2$ upon which f never vanishes. But then, none of f, f'. f'', $f^{(3)}$ would vanish on J_3 contrary to hypothesis.

Let I be an open real interval and let $T_0 = I \setminus S_0$. If $T_0 = \emptyset$, then f, and so $f^{(3)}$, must vanish on I. Otherwise, T_0 is a nonvoid open set and there is an open interval $I_1 \subseteq T_0 \cap I$. Let $T_1 = I_1 \setminus S_1$. Then, as before, either f', and so $f^{(3)}$, vanishes on I_1 or else there is a nonvoid open interval $I_2 \subseteq T_1 \cap I_1 \subseteq I$. Let $T_2 = I_2 \setminus S_2$. Either f'', and so $f^{(3)}$, vanishes on I_2 or there is an nonvoid open subset $I_3 \subseteq T_2 \cap I_2 \subseteq I$. Then either $f^{(3)}$ vanishes on I_3 or there is a nonvoid open interval $I_4 \subseteq T_3 \cap I_3$, where T_3 is the open set $I_3 \setminus S_3$. But this would contradict the density of S_3 in $\mathbb{R}$.

Therefore, $f^{(3)}$ is identically 0 on $\mathbb{R}$, and so therefore, by the Mean Value Theorem, f'' must be constant, f' at most linear and f at most quadratic.

2010:10. Prove that the set $\mathbb{Q}$ of rationals can be written as the union of countably many subsets of $\mathbb{Q}$ each of which is dense in the set $\mathbb{R}$ of real numbers.

Solution. Let $\{r_1, r_2, \ldots, r_i, \ldots\} = \{2, 3, 5, 6, 7, 10, 11, 12, 13, 14, 15, 17, \ldots\}$ be the increasing sequence of all positive integers that are not the mth power of any integer for any integer exponent exceeding 1. Let X_k be the set of all rationals of the form a/r_k^c where a is an integer coprime with r_k and c is a positive integer. We include also in X_1 all integers. Then it can be seen that every rational lies in one of the X_k and that the X_k are disjoint.

Let $k \geq 1$ and $\epsilon > 0$; select c such that $1/r^c < \epsilon$. Then $\{(ar_k+1)/r_k^{-(c+2)} : a \in \mathbb{Z}\}$ is a set of elements of X_k equally spaced at intervals of $1/r_k^{c+1}$ in $\mathbb{R}$, so that an element of the set lies inside each real open interval of length ϵ. Hence X_k is dense in $\mathbb{R}$.

2011:1. Let S be a nonvoid set of real numbers with the property that, for each real number x, there is a unique real number $f(x)$ belonging to S that is farthest from x, i.e., for each y in S distinct from $f(x)$, $|x - f(x)| > |x - y|$. Prove that S must be a singleton.

Solution 1. Let a be an arbitrary real, and let b be the unique point of S that is farthest from a. Then every number in S lies on the same side of b as a. Let c be the unique number in S that is farthest from b. Then S must be in the closed interval bounded by b and c. It follows that both b and c are the farthest points in S from $\frac{1}{2}(b+c)$. By the unicity, we must have that $b = c$ and the result follows.

Solution 2. Since the condition applies to 0 in particular, $|x| \leq |f(0)|$ for all $x \in S$. Therefore S is bounded. Let $a = \inf S$ and $b = \sup S$. Then $a \leq b$ and $S \subseteq [a, b]$. Since $a \leq x \leq f(a) \leq b$, for all $x \in S$, it follows that $f(a) = b$. Similarly $f(b) = a$, and both a and b belong to S.

For each $x \in S$, $|x - \frac{1}{2}(a+b)| \leq |a - \frac{1}{2}(a+b)| = |b - \frac{1}{2}(a+b)|$, so that $a = f(\frac{1}{2}(a+b)) = b$. The desired result follows.

[Problem **2011:1** was contributed by Bamdad Yahaghi.]

2012:6. Find all continuous real-valued functions defined on $\mathbb{R}$ that satisfy $f(0) = 0$ and
$$f(x) - f(y) = (x - y)g(x + y)$$
for some real-valued function $g(x)$.

Solution 1. Setting $y = 0$, we find that $f(x) = xg(x)$ for all real x, so that $g(x)$ is continuous for all nonzero values of x. Therefore

(6.4) $$(x + y)[f(x) - f(y)] = (x - y)f(x + y)$$

for all real x, y. Hence, when $(x + y)(x - y) \neq 0$,

$$\frac{f(x) - f(y)}{x - y} = \frac{f(x + y)}{x + y}.$$

Suppose that $x \neq 0$. Then

$$f'(x) = \lim_{y \to x} \frac{f(x) - f(y)}{x - y} = \frac{f(2x)}{2x}.$$

Taking $y = 1 - x$, we find that

$$f(x) - f(1 - x) = (2x - 1)f(1).$$

Therefore, when $x \neq 0$,

$$f'(x) + f'(1 - x) = 2f(1),$$

so that

$$\frac{f(2x)}{2x} + \frac{f(2 - 2x)}{2(1 - x)} = 2f(1) \Longrightarrow$$
$$(1 - x)f(2x) + xf(2 - 2x) = 4x(1 - x)f(1).$$

Substituting $(2x, 2 - 2x)$ for (x, y) in the Eq. (6.4), we find that

$$f(2x) - f(2 - 2x) = (4x - 2)\frac{f(2)}{2} = (2x - 1)f(2).$$

Hence

$$f(2x) = 4x(1 - x)f(1) + x(2x - 1)f(2)$$

so that

$$f(x) = x(2 - x)f(1) + \frac{1}{2}x(x - 1)f(2) = \left[\frac{f(2)}{2} - f(1)\right]x^2 + \left[2f(1) - \frac{f(2)}{2}\right]x$$

for $x \neq 0$. Since $f(x)$ is continuous, this should hold for all real x, so that $f(x)$ is a linear combination of the functions x and x^2.

It is straightforward to check that $f(x) = x$ and $f(x) = x^2$ satisfy the equation (with $g(x)$ respectively equal to 1 and x), along with any linear combination of these. Hence the general solution is

$$f(x) = ax^2 + bx$$

where a and b are arbitrary constants.

Solution 2. [J. Zung] Since

$$g(x) = \frac{1}{2}\left[f\left(\frac{x}{2} + 1\right) - f\left(\frac{x}{2} - 1\right)\right],$$

we see that $g(x)$ is a continuous function. As in Solution 1, we show that $f'(x) = g(2x)$.

We can verify that the vector space generated by the polynomials x and x^2 consists of solutions of the equation. Given a solution, we can add a polynomial of degree not exceeding 2 to it to get a solution for which $f(-1) = f(0) = f(1) = 0$, so there is no loss of generality in assuming this to be the case.

Since $f(-1) = f(0) = f(1)$, we have that $g(0) = 0$, $f(x) = xg(x)$ and $g(-x) = -g(x)$ for each x. (For the last, observe that for $x \neq 0$, $f(x) - f(-x) = 0$ so that $-xg(-x) = f(-x) = f(x) = xg(x)$.) We have that

$$f(x) - f(1-x) = (2x-1)g(1)$$

and

$$f(x) - f(-1-x) = (2x+1)g(-1) = -(2x+1)g(1),$$

whence

$$f(1-x) - f(-1-x) = -4xg(1).$$

However, from the given equation, we have also that

$$f(1-x) - f(-1-x) = 2g(-2x) = -2g(2x),$$

so that $g(2x) = 2xg(1)$ for all x. Therefore $f(x) = g(1)x^2$. Since $f(1) = 0$, we must have $g(1) = 0$, so that $f(x) \equiv 0$.

The significance of this is that the polynomial we add to a solution f of the equation to get a solution that vanishes at $-1, 0, 1$ is the negative of the solution itself, so that the solution must be itself a polynomial of degree 1 or 2.

2012:7. Consider the following problem:

Suppose that $f(x)$ is a continuous real-valued function defined on the interval $[0, 2]$ for which

$$\int_0^2 f(x)\, dx = \int_0^2 (f(x))^2\, dx.$$

Prove that there exists a number $c \in [0, 2]$ for which either $f(c) = 0$ or $f(c) = 1$.

    (a) Criticize the following solution:

Clearly $\int_0^2 f(x)\, dx \geq 0$. By the Extreme Value Theorem, there exist numbers u and v in $[0, 2]$ for which $f(u) \leq f(x) \leq f(v)$ for $0 \leq x \leq 2$. Hence

$$f(u) \int_0^2 f(x)\, dx \leq \int_0^2 f(x)^2\, dx \leq f(v) \int_0^2 f(x)\, dx.$$

Since $\int_0^2 f(x)^2\, dx = 1 \cdot \int_0^2 f(x)\, dx$, by the Intermediate Value Theorem, there exists a number $c \in [0, 2]$ for which $f(c) = 1$. $\square$

    (b) Show that there is a nontrivial function f that satisfies the conditions of the problem but that never assumes the value 1.

    (c) Provide a complete solution of the problem.

Solution. (a) The solution does not apply when $f(x)$ is identically zero. However, there is a more fundamental difficulty with it. The solution is correct for a nonzero function which is nonnegative at every point of the interval. However, it fails whenever $f(x)$ assumes a negative value. In this case, there are values of x for which $f(u) \leq f(x) < 0$, so that $f(u)f(x) \geq f(x)^2$ and so

$$f(u) \int_N f(x)\, dx \geq \int_N (f(x))^2\, dx$$

where N is the subset of $[0, 2]$ upon which $f(x)$ is negative. This puts into question the displayed inequality in the purported solution.

(b) We observe that when $f(x) = \frac{1}{2}x$, then $\int_0^2 f(x)\,dx > \int_0^2 (f(x))^2\,dx$, while when $f(x) = \frac{1}{2}(x - 2)$, then $\int_0^2 f(x)\,dx < 0 < \int_0^2 (f(x))^2\,dx$. This suggests that we can satisfy the condition of the problem with a function of the form $f(x) = \frac{1}{2}(x - \lambda)$ for some value of λ in the interval $(0, 2)$. Any such function will never assume the value 1 on the interval.

We have that

$$\int_0^2 f(x)\,dx = 1 - \lambda$$

and

$$\int_0^2 (f(x))^2\,dx = \frac{1}{6}[3(1 - \lambda)^2 + 1].$$

These two expressions are equal when $\lambda = \sqrt{2/3}$. Thus, we can take

$$f(x) = \frac{x}{2} - \frac{1}{\sqrt{6}}.$$

Comment. J. Love showed that the function

$$f(x) = \frac{1}{2} - \frac{\sqrt{3}}{4}x = \frac{1}{4}(2 - \sqrt{3}x)$$

satisfies the condition of the problem but does not assume the value 1.

(c) On the one hand, we can complete the proof in (a) by considering the possibility that $f(x)$ takes negative values. Since the integral of $f(x)$ over $[0, 2]$ is nonnegative, $f(x)$ must assume positive values as well. Therefore, by the Intermediate Value Theorem, it must vanish somewhere within the interval.

Alternatively, we can get a more direct argument by considering the function $g(x) = f(x)^2 - f(x)$. By hypothesis, $g(x)$ is a continuous function for which $\int_0^2 g(x)\,dx = 0$. Therefore, either $g(x)$ vanishes identically or it takes both positive and negative values. By the Intermediate Value Theorem, there exists a number c for which $g(c) = 0$, whence $f(c)$ is equal to 0 or 1.

One can also argue by contradiction. Suppose that $f(x)$ takes neither of the values 0 and 1. Then one of the following must occur for each x in the interval $[0, 2]$:

> Case 1: $f(x) < 0$, in which case $f(x) < f(x)^2$;
> Case 2: $0 < f(x) < 1$, in which case $f(x)^2 < f(x)$;
> Case 3: $1 < f(x)$, in which case $f(x) < f(x)^2$.

In Cases 1 and 3, the left side of the given equation is less than the right, while in Case 2, the left side is greater than the right. In all cases, the equality cannot hold, and we obtain the result.

2013:3. Let $f(x)$ be a convex increasing real-valued function defined on the closed interval $[0,1]$ for which $f(0) = 0$ and $f(1) = 1$. Suppose that $0 < a < 1$ and that $b = f(a)$.

    (a) Prove that f is continuous on $(0,1)$.
    (b) Prove that

$$0 \le a - b \le 2 \int_0^1 (x - f(x))\, dx \le 1 - 4b(1 - a).$$

Solution 1. (a) Let $0 \le x < y \le 1$. Then $y = (1 - t)x + t$ where $t = (y - x)/(1 - x)$ and $f(y) \le (1 - t)f(x) + tf(1)$. Therefore

$$0 \le f(y) - f(x) \le t[f(1) - f(x)] = \left[\frac{y - x}{1 - x}\right][1 - f(x)].$$

If follows that $\lim_{y \downarrow x} f(y) = f(x)$ so that f is right continuous at x. A similar argument shows that f is left continuous when $0 < x \le 1$. Therefore f is continuous on $[0,1]$.

    (b) Let $0 < x < 1$. Then $f(x) \le (1 - x)f(0) + xf(1) = x$, so that the graph of $y = f(x)$ over $[0,1]$ lies below the graph of $y = x$. Thus $a \ge b$.

    Let $O = (0,0)$, $I = (1,1)$, $P = (a,b)$ and $Q = (a,a)$. The area between the graphs of $y = f(x)$ and $y = x$ is not less than the combined areas of the triangles OPQ and IPQ, namely $\frac{1}{2}(a - b)a + \frac{1}{2}(a - b)(1 - a) = \frac{1}{2}(a - b)$. Therefore

$$\int_0^1 (x - f(x))\, dx \ge \frac{1}{2}(a - b).$$

    The rightmost inequality is equivalent to $\int_0^1 f(x)\, dx \ge 2b(1 - a)$. We first show that there is a line of equation $y = m(x - a) + b$ passing through (a, b) that passes under the graph of $y = f(x)$.

    Suppose that $0 \le u < x < v \le 1$. Then $(v - u)x = (v - x)u + (x - u)v$ and

$$(v - x)f(x) + (x - u)f(x) = (v - u)f(x) \le (v - x)f(u) + (x - u)f(v).$$

This is equivalent to

$$\frac{f(x) - f(u)}{x - u} \le \frac{f(v) - f(x)}{v - x}.$$

Let m be any number between the supremum of the left side over u and the infimum of the right side over v.

    The line $y = m(x - a) + b$ contains the points $(a - b/m, 0)$ and $(1, m(1 - a) + b)$. By the convexity of f, it cannot contain any point of the graph of

$y = f(x)$ and so lies underneath the graph. The area under the line is equal to

$$\frac{1}{2}\left[\left(1 - a + \frac{b}{m}\right)(m(1 - a) + b)\right] = \frac{2}{m}\left[\frac{m(1-a)+b}{2}\right]^2 \geq 2b(1 - a),$$

by the arithmetic-geometric means inequality. This gives the desired result.

Solution 2 (to part of b). First, we have that $b = f(a) \leq (1 - a)f(0) + af(1) = a$, so that $a - b \geq 0$. Making use of the respective substitutions $x = ta$ and $x = (1 - t)a + t = a + t(1 - a)$ we find that

$$\int_0^a (x - f(x))\, dx = a\int_0^1 (ta - f(ta))\, dt \geq a\int_0^1 (ta - tf(a))\, dt$$

$$= a\int_0^1 t(a - b)\, dt = \frac{a(a - b)}{2},$$

and that

$$\int_a^1 (x - f(x))\, dx = (1 - a)\int_0^1 [(1 - t)a + t - f((1 - t)a + t)]\, dt$$

$$\geq (1 - a)\int_0^1 [(1 - t)a + t - ((1 - t)f(a) + tf(1))]\, dt$$

$$= (1 - a)\int_0^1 (1 - t)(a - b)\, dt = \frac{(1 - a)(a - b)}{2}.$$

Adding these two inequalities together yields

$$\int_0^1 (x - f(x))\, dx \geq \frac{a - b}{2}.$$

2013:10. (a) Let f be a real-valued function defined on the real number field $\mathbf{R}$ for which $|f(x) - f(y)| < |x - y|$ for any pair (x, y) of distinct elements of $\mathbf{R}$. Let $f^{(n)}$ denote the nth composite of f defined by $f^{(1)}(x) = f(x)$ and $f^{(n+1)}(x) = f(f^{(n)}(x))$ for $n \geq 2$. Prove that exactly one of the following situations must occur:

   (i) $\lim_{n \to +\infty} f^{(n)}(x) = +\infty$ for each real x;
   (ii) $\lim_{n \to +\infty} f^{(n)}(x) = -\infty$ for each real x;
   (iii) there is a real number z such that

$$\lim_{n \to +\infty} f^{(n)}(x) = z$$

   for each real x.

   (b) Give examples to show that each of the three cases in (a) can occur.

Solution. (a) Note that the condition on f implies that f is continuous. Suppose that there exists a real number z for which $f(z) = z$. Let $x \neq z$. Then $|f(x) - f(z)| = |f(x) - z| < |x - z|$, so that $f(x) \neq x$. Define $x_n = f^{(n)}(x)$ and let u be the limit of the decreasing sequence $\{|x_n - z|\}$. Note that either $x_n \leq z - u$ or $x_n > z + u$ for each n, so that as n increases, x_n is either arbitrarily close to $z - u$ on the left or arbitrarily close to $z + u$ on the right.

If $\lim_{n\to\infty} x_n = z - u$, then

$$z - u = \lim_{n\to\infty} f^{(n+1)}(x_n) = \lim_{n\to\infty} f(f^n(x_n)) = f(z - u)$$

from which $u = 0$ and $\lim_{n\to\infty} x_n = z$. Similarly, if $\lim_{n\to\infty} x_n = z + u$, then $u = 0$ and $\lim_{n\to\infty} x_n = z$.

Otherwise, given $\epsilon > 0$, we can find a positive integer n for which $z - u - \frac{1}{2}\epsilon < x_n \leq z - u$ and $z + u \leq x_{n+1} < z + u + \frac{1}{2}\epsilon$. Therefore

$$|f(z - u) - x_{n+1}| = |f(z - u) - f(x_n)| < |z - u - x_n| < \frac{1}{2}\epsilon,$$

from which

$$z + u - \epsilon < x_{n+1} - \frac{1}{2}\epsilon < f(z - u) < x_{n+1} + \frac{1}{2}\epsilon < z + u + \epsilon.$$

Since this holds for all positive ϵ, $f(z - u) = f(z + u)$. Similarly, $f(z + u) = f(z - u)$. Hence

$$2u = |(z + u) - (z - u)| = |f(z - u) - f(z + u)| = |f(z + u) - f(z - u),$$

so that $u = 0$ and $\lim_{n\to\infty} x_n = z$.

Suppose that there is no z for which $f(z) = z$. Then the function $g(x) = f(x) - x$ is continuous and vanishes nowhere on $\mathbb{R}$. If $g(x) > 0$ for all x, then for each x, $f^{(n)}(x)$ is an increasing sequence. It cannot have a finite limit, since this limit would be a fixpoint of f. Therefore the sequence must diverge to infinity. The case that $g(x) < 0$ for all x can be similarly handled.

(b) (i) Let $f(x) = (1 + x^2)^{1/2}$. Then $f(x) > x$ and so $f^{(n)}(x)$ is a strictly increasing function of n. Since $f'(x) = x(1 + x^2)^{-1/2}$, we find that, given x, y, by the Mean Value Theorem, there exists a number z for which

$$|f(x) - f(y)| = |f'(z)||x - y| < |x - y|.$$

(ii) Let $f(x) = -(1 + x^2)^{1/2}$.
(iii) Let $f(x) = \frac{1}{2}x$.

2014:4. Let $f(x)$ be a continuous real-valued function on $[0, 1]$ for which

$$\int_0^1 f(x)dx = 0 \quad \text{and} \quad \int_0^1 xf(x)\, dx = 1.$$

(a) Give an example of such a function.
(b) Prove that there is a nontrivial open interval I contained in $(0, 1)$ for which $|f(x)| > 4$ for $x \in I$.

Solution. (a) Trying a function of the form $c(x - \frac{1}{2})$ yields the examples $f(x) = 12x - 6$. Other examples are $f(x) = -\frac{1}{2}\pi^2 \cos(\pi x)$ and $f(x) = -2\pi \sin(2\pi x)$.

(b) If there existed no such interval I, then $|f(x)| \leq 4$ for all $x \in [0, 1]$. Then

$$1 = \int_0^1 f(x)(x-(1/2))\,dx = \int_0^{\frac{1}{2}} (-f(x))((1/2)-x)\,dx + \int_{\frac{1}{2}}^1 f(x)(x-(1/2))\,dx$$

$$< 4\int_0^{\frac{1}{2}} ((1/2) - x)dx + 4\int_{\frac{1}{2}}^1 (x - (1/2))\,dx = 4\cdot(1/8) + 4\cdot(1/8) = 1.$$

Note that the inequality is strict. Equality would require that $f(x) = -4$ on $(0, \frac{1}{2})$ and $f(x) = 4$ on $(\frac{1}{2}, 1)$, an impossibility for a continuous function.

Comment. B. Yaghani draws attention to Problem 25 of Chap. 11 appearing on page 184 of Michael Spivak, *Calculus* (W.A. Benjamin, 1967), where it is required to show that, if g is twice differentiable on $[0, 1]$ with $g(0) = g'(0) = g'(1) = 0$ and $g(1) = 1$, then $|g''(c)| \geq 4$ for some $c \in [0, 1]$. Setting $g(t) = \int_0^t (x - t)f(x)dx$, whereupon $g'(t) = -\int_0^t f(x)dx$ and $g''(t) = -f(t)$, yields the desired result.

[Problem **2014:4** is based on #4520 in the *American Mathematical Monthly* 60:1 (January, 1953), 48; 61:4 (April, 1954), 266–268.]

2014:10. Does there exist a continuous real-valued function defined on **R** for which $f(f(x)) = -x$ for all $x \in \mathbb{R}$?

Solution 1. The answer is **no**. Since $f(f(x)) = -x$ defines a one-one function, it follows that $f(x)$ itself is one-one. Therefore $f(x)$ is either strictly increasing or strictly decreasing on $\mathbb{R}$. But this is impossible, since in either case, $f(f(x))$ would be increasing.

Solution 2. The answer is **no**. Suppose that f is such a function and that $f(0) = u$. Then $f(u) = f(f(0)) = 0$ and so $u = f(0) = f(f(u)) = -u$. Hence $u = 0$ and $f(0) = 0$.

Suppose that $f(v) = 0$. Then $f(0) = f(f(v)) = -v$. Therefore $f(v) = 0$ if and only if $v = 0$.

Let $a \neq 0$ and $b = f(a)$. Then $f(b) = -a$ and $f(-a) = -b$. Consider the four numbers $a, b, -a, -b$ repeated cyclically. The alternate ones have opposite signs, so there must be two numbers of the same sign followed by one of the opposite sign. With no loss of generality, we can assume that a and b are positive while $-a$ and $-b$ are negative. Then $f(a)$ and $f(b)$ have opposite signs, so by the Intermediate Value Theorem, there must exist c between a and b for which $f(c) = 0$. But this is a contradiction. Therefore, there is no continuous function with the desired property.

2015:1. Suppose that u and v are two real-valued functions defined on the set of reals. Let $f(x) = u(v(x))$ and $g(x) = u(-v(x))$ for each real x. If $f(x)$ is continuous, must $g(x)$ also be continuous?

Solution 1. The answer is **no**. Let

$$v(x) = \begin{cases} 0, & \text{when } x \leq 0; \\ 1, & \text{when } x > 0. \end{cases}$$

Suppose that u is any function for which $u(-1) = 1$ and $u(0) = u(1) = 0$. Then $f(x) = 0$ for all real x while

$$g(x) = \begin{cases} 0, & \text{when } x \leq 0; \\ 1, & \text{when } x > 0. \end{cases}$$

Solution 2. (Xuan Tu) An example for which $g(x)$ fails to be continuous at -1 is $(u(x), v(x)) = ((1 + x)^{-1}, x^2)$.

2015:3. Let $f : [0, 1] \longrightarrow \mathbb{R}$ be continuously differentiable. Prove that

$$\left| \frac{f(0) + f(1)}{2} - \int_0^1 f(x)\, dx \right| \leq \frac{1}{4} \sup\{|f'(x)| : 0 \leq x \leq 1\}.$$

Solution 1. Integrating by parts, we find that

$$\int_0^1 \left(x - \frac{1}{2} \right) f'(x)\, dx = \left[\left(x - \frac{1}{2} \right) f(x) \right]_0^1 - \int_0^1 f(x)\, dx$$

$$= \frac{f(1) + f(0)}{2} - \int_0^1 f(x)\, dx.$$

Since

$$\left| \int_0^1 \left(x - \frac{1}{2} \right) f'(x)\, dx \right| \leq \sup |f'(x)| \int_0^1 \left| x - \frac{1}{2} \right| dx = \frac{1}{4} \sup |f'(x)|,$$

the desired result follows.

Solution 2. Note that

$$\frac{f(1) - f(0)}{2} - \int_0^1 (f(x) - f(0))\, dx = \frac{f(0) + f(1)}{2} - \int_0^1 f(x)\, dx.$$

Because of this relation and because $f(t) - f(0)$ and $f(t)$ have the same derivative, there is no loss in generality in assuming that $f(0) = 0$. Then

$$\frac{f(1)}{2} - \int_0^1 f(x)\, dx = \frac{1}{2} \int_0^1 f'(x)\, dx - \int_0^1 \int_0^x f'(y)\, dy\, dx$$

$$= \frac{1}{2} \int_0^1 f'(y)\, dy - \int_0^1 f'(y) \int_y^1 dx\, dy$$

$$= \frac{1}{2} \int_0^1 f'(y) \, dy - \int_0^1 f'(y)(1-y) \, dy$$

$$= \int_0^1 \left(y - \frac{1}{2} \right) f'(y) \, dy$$

and we can conclude as in Solution 1.

Comment. A weaker result can be obtained by noting that the Integral Mean Value Theorem provides a value $c \in [0, 1]$ for which $f(c) = \int_0^1 f(x) dx$. The left side is thus equal to

$$\frac{1}{2} |(f(1) - f(c)) + (f(0) - f(c))| \leq \frac{1}{2} (|f(1) - f(c)| + |f(c) - f(0)|)$$

$$= \frac{1}{2} \left(\left| \int_c^1 f'(x) \, dx \right| + \left| \int_0^c f'(x) \, dx \right| \right)$$

$$\leq \int_0^1 |f'(x)| \, dx \leq \frac{1}{2} \sup |f'(x)|.$$

[Problem **2015:3** was contributed by Omran Kouba, Higher Institute for Applied Sciences and Technology, Damascus, Syria.]

CHAPTER 7

Linear Algebra

2001:4. Let V be the vector space of all continuous real-valued functions defined on the open interval $(-\pi/2, \pi/2)$, with the sum of two functions and the product of a function and a real scalar defined in the usual way.

   (a) Prove that the set $\{\sin x, \cos x, \tan x, \sec x\}$ is linearly independent.

   (b) Let W be the linear space generated by the four trigonometric functions given in (a), and let T be the linear transformation determined on W into V by $T(\sin x) = \sin^2 x$, $T(\cos x) = \cos^2 x$, $T(\tan x) = \tan^2 x$ and $T(\sec x) = \sec^2 x$. Determine a basis for the kernel of T.

Solution 1.

   (a) Suppose that $a \sin x + b \cos x + c \tan x + d \sec x = 0$ identically for all $x \in (-\frac{\pi}{2}, \frac{\pi}{2})$. Then evaluating this equation at $0, \pi/6, \pi/4, \pi/3$ leads to a system of four linear equations in a, b, c, d whose sole solution is given by $a = b = c = d = 0$.

   (b) Since $\sin^2 x + \cos^2 x = 1$ and $1 + \tan^2 x = \sec^2 x$, it is clear from the linearity of T that $T(\sin x + \cos x + \tan x - \sec x) = 0$. On the other hand, by evaluating at three of the values of x used in (a), we can see that $\{\sin^2 x, \cos^2 x, \tan^2 x\}$ is linearly independent. Thus the nullity of T is at least 1 and the rank of T is at least 3. Hence the kernel of T has dimension 1 and must be the span of $\{\sin x + \cos x + \tan x - \sec x\}$.

Solution 2.

   (a)

$$a \sin x + b \cos x + c \tan x + d \sec x = 0 \ (\forall x)$$

$$\Leftrightarrow a \sin x \cos x + b \cos^2 x + c \sin x + d = 0 \ (\forall x)$$

$$\Leftrightarrow \sin x(a \cos x + c) = -(b \cos^2 x + d) \ (\forall x)$$

$$\Rightarrow (1 - \cos^2 x)(a^2 \cos^2 x + 2ac \cos x + c^2) = b^2 \cos^4 x + 2bd \cos^2 x + d^2 \ (\forall x)$$

$$\Leftrightarrow (b^2 + a^2) \cos^4 x + 2ac \cos^3 x + (c^2 - a^2 + 2bd) \cos^2 x - 2ac \cos x + (d^2 - c^2) = 0 \ (\forall x)$$

$$\Leftrightarrow a = b = c = d = 0,$$

© Springer International Publishing Switzerland 2016   107
E.J. Barbeau, *University of Toronto Mathematics Competition (2001–2015)*, Problem Books in Mathematics,
DOI 10.1007/978-3-319-28106-3_7

from which the linear independence follows. (Note that the left expression is a polynomial in $\cos x$ which vanishes at infinitely many values.)

(b) $a \sin x + b \cos x + c \tan x + d \sec x$ belongs to the kernel of T

$$\Leftrightarrow a \sin^2 x + b \cos^2 x + c \tan^2 x + d \sec^2 x = 0 \ (\forall x)$$

$$\Leftrightarrow a + (b - a) \cos^2 x + (c + d) \sec^2 x - c = 0 \ (\forall x)$$

$$\Leftrightarrow (a - c) + (b - a) \cos^2 x + (c + d) \sec^2 x = 0 \ (\forall x)$$

$$\Leftrightarrow (b - a) \cos^4 x + (a - c) \cos^2 x + (c + d) = 0 \ (\forall x)$$

$$\Leftrightarrow b = a = c = -d,$$

which occurs if and only if $a \sin x + b \cos x + c \tan x + d \sec x$ is a multiple of $\sin x + \cos x + \tan x - \sec x$. Hence the nullity of T is 1 and the rank of T is 3. A basis for the kernel of T consists of the function $\sin x + \cos x + \tan x - \sec x$.

2002:7. Prove that no vector space over $\mathbb{R}$ is a finite union of proper subspaces.

Solution. Suppose that the statement is false. Let r be the minimum number of proper subspaces whose union is the whole of V. Suppose that $S_1, S_2, \ldots, S_r$ are subspaces for which $V = \cup_{i=1}^r S_i$. Note that $r \geq 2$. Since r is minimal, no subspace is contained in a union of the others. Select $v \in S_1 \setminus \cup_{i=2}^r S_i$ and $w \in S_2 \setminus S_1$. For each real λ, $\lambda v + w \notin S_1$. By the pigeonhole principle, there is an index j and two distinct reals μ and ν for which $\mu v + w$ and $\nu v + w$ both belong to S_j. Hence, $(\mu - \nu)v \in S_j$, so that $v \in S_j$. But this contradicts the choice of v.

[Problem **2002:7** is #10707 in the *American Mathematical Monthly* 106:1 (January, 1999), 67; 107:10 (December, 2000), 951–952.]

2002:8. [(a) Suppose that P is an $n \times n$ nonsingular matrix and that u and v are column vectors with n components. The matrix $v^t P^{-1} u$ is 1×1, and so can be identified with a scalar. Suppose that its value is not equal to -1. Prove that the matrix $P + uv^t$ is nonsingular and that

$$(P + uv^t)^{-1} = P^{-1} - \frac{1}{\alpha} P^{-1} uv^t P^{-1}$$

where v^t denotes the transpose of v and $\alpha = 1 + v^t P^{-1} u$. (b) Explain the situation when $\alpha = 0$.

Solution 1. (a) Observe that uv^t is an $n \times n$ matrix and that for each column vector w, $v^t w$ is a 1×1 matrix or scalar. Thus, $(uv^t)w = u(v^t w) = (v^t w)u$, and so uv^t is a rank 1 matrix, whose range is spanned by the vector u. It follows that the matrix $uv^t P^{-1}$ is also rank 1 whose range is the span of u. Note that, for every real scalar λ,

$$(uv^t P^{-1})\lambda u = \lambda u(v^t P^{-1} u) = \lambda (v^t P^{-1} u)u$$

since $v^t P^{-1} u$, being a 1×1 matrix, is essentially a scalar. Thus, on the span of u, $(uv^t P^{-1})$ behaves like $(v^t P^{-1} u)I$, a multiple of the identity. It follows that

$$(v^t P^{-1} u I - uv^t P^{-1})(uv^t P^{-1}) = O.$$

Now we are ready to establish the result.

$$(P + uv^{t})(P^{-1} - \frac{1}{\alpha}P^{-1}uv^{t}P^{-1})$$

$$= I + uv^{t}P^{-1} - \frac{1}{\alpha}uv^{t}P^{-1} - \frac{1}{\alpha}uv^{t}P^{-1}uv^{t}P^{-1}$$

$$= I + \frac{1}{\alpha}[(\alpha - 1)uv^{t}P^{-1} - uv^{t}P^{-1}uv^{t}P^{-1}]$$

$$= I + \frac{1}{\alpha}[(v^{t}P^{-1}u)uv^{t}P^{-1} - uv^{t}P^{-1}uv^{t}P^{-1}]$$

$$= I + \frac{1}{\alpha}[(v^{t}P^{-1}uI - uv^{t}P^{-1})(uv^{t}P^{-1})] = I,$$

as desired.

(b) Suppose that $v^{t}P^{-1}u = -1$. Then $v^{t}P^{-1}uv^{t} = -v^{t} = -v^{t}P^{-1}P$, so that $v^{t}P^{-1}(uv^{t} + P) = O$. Since $v^{t}P^{-1} \neq O$, we cannot have an inverse for $uv^{t} + P$.

Solution 2. This is similar to Solution 1, with the inverse checked on the right instead of the left. Let $z = P^{-1}u$. Since uv^{t} is a rank 1 matrix whose range is spanned by u (see Solution 1), $P^{-1}uv^{t}$ is a rank 1 matrix whose range is spanned by z. We have that

$$(P^{-1}uv^{t})\lambda z = \lambda(P^{-1}uv^{t}P^{-1})u = \lambda(P^{-1}u)(v^{t}P^{-1}u)$$
$$= \lambda(v^{t}P^{-1}u)(P^{-1}u) = \lambda(v^{t}P^{-1}u)z$$

so that, on the span of z, $P^{-1}uv^{t}$ behaves like $(v^{t}P^{-1}u)I$. Thus

$$(v^{t}P^{-1}uI - P^{-1}uv^{t})(P^{-1}uv^{t}) = 0.$$

Note that

$$(P^{-1} - \frac{1}{\alpha}P^{-1}uv^{t}P^{-1})(P + uv^{t})$$

$$= I + P^{-1}uv^{t} - \frac{1}{\alpha}P^{-1}uv^{t} - \frac{1}{\alpha}P^{-1}uv^{t}P^{-1}uv^{t}$$

$$= I + \frac{1}{\alpha}[(\alpha - 1)P^{-1}uv^{t} - P^{-1}uv^{t}P^{-1}uv^{t}]$$

$$= I + \frac{1}{\alpha}[(v^{t}P^{-1}u)P^{-1}uv^{t} - P^{-1}uv^{t}P^{-1}uv^{t}]$$

$$= I + \frac{1}{\alpha}[(v^{t}P^{-1}uI - P^{-1}uv^{t})P^{-1}uv^{t}]$$

from which the result follows.

Comment. This is known as the *Sherman-Morrison Formula* (see G.H.Golub, C.F.Van Loon, *Matrix computation*, third edition, 1996, page 51).

2003:8. Three $m \times n$ matrices A, B and $A + B$ have rank 1. Prove that either all the rows of A and B are multiples of one and the same vector, or that all of the columns of A and B are multiples of one and the same vector.

Solution 1. Let $A = (a_{ij})$ and $B = (b_{ij})$. Since the image each of the matrix A and the matrix B as an operator on $\mathbb{F}^n$ (where $\mathbb{F}$ is the underlying field) is the span of its column vectors, there exist nonzero vectors (a_i), (b_i), (u_j), (v_j) for which

$$a_{ij} = a_i u_j \text{ and } b_{ij} = b_i v_j.$$

Hence $a_{ij} + b_{ij} = a_i u_j + b_i v_j$. Similarly, since $A + B = (a_{ij} + b_{ij})$ has rank 1, there are vectors (c_i) and (w_j) for which

$$a_{ij} + b_{ij} = c_i w_j.$$

Thus, for all i and j,

$$a_i u_j + b_i v_j = c_i w_j.$$

If the rows of A and B are multiples of the same row, then there exists a constant k for which $v_j = k u_j$ for each j, and so $c_i w_j = (a_i + k b_i) u_j$, and so the ith row of $A + B$ is also a multiple of (u_j).

Suppose that the vectors $(u_1, \ldots, u_j, \ldots, u_n)$ and $(v_1, \ldots, v_j, \ldots, v_n)$ are linearly independent. Wolog, we may assume that (u_1, u_2) and (v_1, v_2) are linearly independent. Then $u_1 v_2 - u_2 v_1 \neq 0$, and for each i, the system

$$u_1 a_i + v_1 b_i = w_1 c_i$$
$$u_2 a_i + v_2 b_i = w_2 c_i$$

has the solution $a_i = p c_i$, $b_i = q c_i$, where

$$p = \frac{w_1 v_2 - w_2 v_1}{u_1 v_2 - u_2 v_1}$$

and

$$q = \frac{u_1 w_2 - u_2 w_1}{u_1 v_2 - u_2 v_1}$$

are independent of i. The result follows.

Solution 2. Suppose that $A = \mathbf{u} \mathbf{a}^t$ and $B = \mathbf{v} \mathbf{b}^t$ where $\mathbf{a} = (a_j)$, $\mathbf{b} = (b_j)$, $\mathbf{u} = (u_i)$ and $\mathbf{v} = (v_i)$ are column vectors. If the set $\{\mathbf{u}, \mathbf{v}\}$ is linearly dependent, then the columns of $A + B$, A and B are multiplies of each other and the result follows. Otherwise, let $\{\mathbf{u}, \mathbf{v}\}$ be linearly independent. The ith columns of A and B are respectively $a_i \mathbf{u}$ and $b_i \mathbf{v}$. Since $A + B$ is of rank 1, there exists a column vector $\mathbf{w}$ and scalars c_i for which

$$a_i \mathbf{u} + b_i \mathbf{v} = c_i \mathbf{w}$$

for each i. If $c_i = 0$, then $a_i = b_i = 0$. Since not all c_i vanish, we may suppose wolog that $c_1 \neq 0$ and that $a_1 \neq 0$.

For each i, j,

$$c_i(a_j\mathbf{u} + b_j\mathbf{v}) - c_j(a_i\mathbf{u} + b_i\mathbf{v}) = c_i(c_j\mathbf{w}) - c_j(c_i\mathbf{w}) = \mathbf{0}$$
$$\implies (c_i a_j - c_j a_i)\mathbf{u} + (c_i b_j - c_j b_i)\mathbf{v} = \mathbf{0}$$
$$\implies c_i a_j - c_j a_i = c_i b_j - c_j b_i = 0$$
$$\implies c_i a_j = c_j a_i \text{ and } c_i b_j = c_j b_i$$
$$\implies c_i c_j(a_j b_i - a_i b_j) = 0.$$

If $c_i \neq 0$, then taking $j = 1$, we find that $b_i = (a_i/a_1)b_1$ as well as $a_i = (a_i/a_1)a_1$. Since this is true for $i = 1$ and when $c_i \neq 0$, we conclude that the vectors $\mathbf{b}$ and $\mathbf{a}$ are parallel, and the rows of $A + B$ are sums of multiples of these parallel vectors. The result follows.

2004:5. Let A be a $n \times n$ matrix with determinant equal to 1. Let B be the matrix obtained by adding 1 to every entry of A. Prove that the determinant of B is equal to $1 + s$, where s is the sum of the n^2 entries of A^{-1}.

Solution 1. First, we make a general observation. Let $U = (u_{ij})$ and $V = (v_{ij})$ be two $n \times n$ matrices. Then $\det(U + V)$ is the sum of the determinants of 2^n $n \times n$ matrices W_S, where S is a subset of $\{1, 2, \ldots n\}$ and the (i, j)th element of W_S is equal to u_{ij} when $i \in S$ and v_{ij} when $i \notin S$. In the special case that $U = A$ and $V = E$, the matrix whose every entry is equal to 1, W_S is equal to 0 except when $S = \{1, 2, \ldots, n\}$ or $S = \{1, 2, \ldots, n\} \setminus \{k\}$ for some integer k. In the former case, $\det W_S = \det A = 1$, and in the latter case, all rows of S except the kth agree with the corresponding rows of A and the kth row of W_S consists solely of 1s, so that the determinant of S is equal to $A_{k,1} + A_{k,2} + \cdots + A_{k,n}$, where A_{ij} is the cofactor of a_{ij} in the expansion of $\det A$. Thus,

$$\det(A + E) = \det A + \sum_{k=1}^{n}\left(\sum_{l=1}^{n} A_{kl}\right) = 1 + \sum_{i,j} A_{ij},$$

as desired.

Solution 2. [R. Barrington Leigh] Let $A^{-1} = (c_{ij})$ and $d_j = \sum_{i=1}^{n} c_{ij}$, the jth column sum of A^{-1} for $1 \leq j \leq n$. Let E be the $n \times n$ matrix all of whose entries are 1, so that $B = A + E$. Observe that $E = (1, 1, \ldots, 1)^{\mathbf{t}}(1, 1, \ldots, 1)$, where $\mathbf{t}$ denotes the transpose. Then

$$\det B = (\det B)(\det A)^{-1} = (\det B)(\det A^{-1}) = \det(BA^{-1})$$
$$= \det[(A + E)A^{-1}] = \det[I + (1, 1, \ldots, 1)^{\mathbf{t}}(1, 1, \ldots, 1)(c_{ij})]$$
$$= \det[I + (1, 1, \ldots, 1)^{\mathbf{t}}(d_1, d_2, \ldots, d_n)].$$

Thus, we need to calculate the determinant of the matrix

$$\begin{pmatrix} 1+d_1 & d_2 & \cdots & d_n \\ d_1 & 1+d_2 & \cdots & d_n \\ & & \cdots & \\ d_1 & d_2 & \cdots & 1+d_n \end{pmatrix}.$$

This is equal to the determinant of the matrix

$$\begin{pmatrix} 1+d_1 & d_2 & \cdots & d_n \\ -1 & 1 & \cdots & 0 \\ & & \cdots & \\ -1 & 0 & \cdots & 1 \end{pmatrix}.$$

which in turn is equal to the determinant of a matrix whose top row is $(1+d_1+d_2+\cdots+d_n, 0, 0, \ldots, 0)$ and for which the cofactor of the top left element is the $(n-1) \times (n-1)$ identity matrix. The result follows.

Solution 3. [M.-D. Choi] *Lemma:* If C is a rank 1 $n \times n$ matrix, then the determinant of $I+C$ is equal to $1 + \text{trace } C$.

Proof of Lemma. The sum of the eigenvalues of C is equal to the trace of C (sum of the diagonal elements). Since 0 is an eigenvalue of algebraic multiplicity $n-1$, the remaining eigenvalue is equal to the trace of C. The eigenvalues of $I+C$ are 1 with algebraic multiplicity $n-1$ and $1 + \text{trace } C$, so that the determinant of $I+C$, which is the product of its eigenvalues, is equal to $1 + \text{trace } C$. $\square$

In the problem, with E defined as in the previous solutions,

$$\det(A+E) = \det(A) \ \det(I+A^{-1}E) = \det(I+A^{-1}E).$$

Since E is of rank 1, so is $A^{-1}E$. The diagonal elements of $A^{-1}E$ are the row sums of A^{-1} and so the trace of $A^{-1}E$ is equal to the sum of all the elements of A^{-1}. The desired result follows.

2004:8. Let V be a complex n-dimensional inner product space. Prove that

$$|u|^2|v|^2 - \frac{1}{4}|u-v|^2|u+v|^2 \le |(u,v)|^2 \le |u|^2|v|^2.$$

Solution. The right inequality is the Cauchy-Schwarz Inequality. We have that

$$4|(u,v)|^2 - 4|u|^2|v|^2 + |u-v|^2|u+v|^2$$
$$= 4(u,v)(v,u) - 4(u,u)(v,v) + (u-v,u-v)(u+v,u+v)$$
$$= 4(u,v)(v,u) - 4(u,u)(v,v) + [\overline{(u,u)+(v,v)} - \overline{(u,v)+(v,u)}]$$
$$\times [(u,u)+(v,v) + (u,v)+(v,u)]$$
$$= 4(u,v)(v,u) - 4(u,u)(v,v) + [(u,u)^2 + (v,v)^2 + 2(u,u)(v,v)]$$
$$- [(u,v)^2 + 2(u,v)(v,u) + (v,u)^2]$$

$$= [(u, u) - (v, v)]^2 - [(u, v) - (v, u)]^2$$
$$= [|u|^2 - |v|^2]^2 - [2i \text{ Im } (u, v)]^2$$
$$= [|u|^2 - |v|^2]^2 + 4[\text{Im } (u, v)]^2 \geq 0.$$

[Problem **2004:8** was posed by Peter Rosenthal in a problems newsletter in the University of Toronto Mathematics Department in 1967.]

2005:3. How many $n \times n$ invertible matrices A are there for which all the entries of both A and A^{-1} are either 0 or 1?

Solution 1. Let $A = (a_{ij})$ and $A^{-1} = (b_{ij})$. Since A and A^{-1} are both invertible, there is at least one 1 in each row and in each column of either of these matrices. Suppose that $1 \leq k \leq n$. Since $1 = \sum_{j=1}^n a_{kj}b_{jk}$, there is a unique index m for which $a_{km} = b_{mk} = 1$. If $l \neq k$, then $0 = \sum_{j=1}^n a_{lj}b_{jk} = \sum_{j=1}^n a_{kj}b_{jl}$, so that $a_{lm} = b_{ml} = 0$. Thus, the mapping $k \longrightarrow m$ is one-one and each row and each column of each matrix contains exactly one 1. Thus A is a permutation matrix (as is A^{-1}), and there are $n!$ possibilities.

Solution 2. Let $\mathbf{e} = (1, 1, 1, \ldots, 1)^t$. Then each component of $A\mathbf{e}$ is a positive integer. The kth component of $\mathbf{e}$ is the kth component of $A^{-1}(A\mathbf{e})$, namely $\sum_{j=1}^n b_{kj}(A\mathbf{e})_j$, where b_{ij} is the (i, j)th component of A^{-1}. Thus

$$1 = \sum_{j=1}^n b_{kj}(A\mathbf{e})_j$$

so that there is at most one value of i for which $b_{ki} \neq 0$ and $b_{ki} = (A\mathbf{e})_i = 1$. Since A^{-1} is invertible, there is at least one 1 in each row, so there must be exactly one in each row. Similarly, there is at least one 1 in each column, so that A^{-1} must be a permutation matrix, as must be A. Therefore, there are exactly $n!$ possibilities.

Solution 3. [E. Redelmeier] Let $A = (a_{ij})$ and $A^{-1} = (b_{ij})$. Consider any value of i. Since $1 = \sum_{j=1}^n a_{ij}b_{ji}$, there is a unique value of j, say $j = \sigma(i)$ for which $a_{ij} = b_{ji} = 1$. Suppose, for $i \neq l$, that $\sigma(i) = \sigma(l)$. Then $0 = \sum_{j=1}^n a_{ij}b_{jl} \geq a_{i\sigma(i)}b_{\sigma(l)l} = 1$, a contradiction. Hence σ is one-one and so a permutation.

Suppose, for $k \neq \sigma(i)$, $a_{ik} = a_{i\sigma(i)}$. Then $k = \sigma(l)$, for some index l distinct from i and so $\sum_{j=1}^n a_{ij}b_{jl} \geq a_{ik}b_{kl} \geq a_{i\sigma(i)}b_{\sigma(l)l} = 1$, a contradiction. Hence each row contains exactly one 1 and, moreover, A is a permutation matrix. So also is A^{-1}. Conversely, each permutation matrix has the desired property, and so the number of matrices is $n!$.

2005:4. Let a be a nonzero real and $\mathbf{u}$ and $\mathbf{v}$ be real 3-vectors. Solve the equation

$$2a\mathbf{x} + (\mathbf{v} \times \mathbf{x}) + \mathbf{u} = \mathbf{O}$$

for the vector $\mathbf{x}$.

Solution 1. If $\mathbf{v} = \mathbf{0}$, then $\mathbf{x} = (2a)^{-1}\mathbf{u}$. Let $\mathbf{v} \neq 0$. Take the dot and cross products by $\mathbf{v}$ to get

$$2a(\mathbf{v} \cdot \mathbf{x}) + \mathbf{v} \cdot \mathbf{u} = 0$$

and

$$2a(\mathbf{v} \times \mathbf{x}) + \mathbf{v} \times (\mathbf{v} \times \mathbf{x}) + \mathbf{v} \times \mathbf{u} = \mathbf{O}.$$

The second equation can be manipulated to

$$2a(\mathbf{v} \times \mathbf{x}) + (\mathbf{v} \cdot \mathbf{x})\mathbf{v} - (\mathbf{v} \cdot \mathbf{v})\mathbf{x} + (\mathbf{v} \times \mathbf{u}) = \mathbf{O}$$

whence

$$\mathbf{v} \times \mathbf{x} = -\frac{1}{2a}[(\mathbf{v} \cdot \mathbf{x})\mathbf{v} - (\mathbf{v} \cdot \mathbf{v})\mathbf{x} + (\mathbf{v} \times \mathbf{u})]$$

$$= \frac{1}{4a^2}(\mathbf{v} \cdot \mathbf{u})\mathbf{v} + \frac{1}{2a}(\mathbf{v} \cdot \mathbf{v})\mathbf{x} - \frac{1}{2a}(\mathbf{v} \times \mathbf{u}).$$

Therefore

$$2a\mathbf{x} + \mathbf{u} = -\frac{1}{4a^2}(\mathbf{v} \cdot \mathbf{u})\mathbf{v} - \frac{1}{2a}(\mathbf{v} \cdot \mathbf{v})\mathbf{x} + \frac{1}{2a}(\mathbf{v} \times \mathbf{u})$$

which can be rearranged to give

$$[8a^3 + 2a(\mathbf{v} \cdot \mathbf{v})]\mathbf{x} = -(\mathbf{v} \cdot \mathbf{u})\mathbf{v} + 2a(\mathbf{v} \times \mathbf{u}) - 4a^2\mathbf{u}.$$

Observe that, since $\mathbf{v} \cdot \mathbf{v} > 0$, the coefficient of $\mathbf{x}$ is nonzero and the same sign as a.

Solution 2. Suppose that $\mathbf{v} = \lambda\mathbf{u}$ for some real λ. Then, since $\mathbf{v} \times \mathbf{x}$ is orthogonal to both $\mathbf{v}$ and $\mathbf{x}$, we must have that $\mathbf{x} = (2a)^{-1}\mathbf{u}$.

Suppose that $\mathbf{v}$ is not a multiple of $\mathbf{u}$. Then we can write $\mathbf{u} = \lambda\mathbf{v} + \mathbf{w}$ where $\mathbf{w} \neq \mathbf{O}$ and $\mathbf{v} \perp \mathbf{w}$. Then $\{\mathbf{v}, \mathbf{w}, \mathbf{v} \times \mathbf{w}\}$ is an orthogonal basis for 3-space. Define the operator

$$T(\mathbf{x}) = 2a\mathbf{x} + (\mathbf{v} \times \mathbf{x}).$$

Then $T(\mathbf{v}) = 2a\mathbf{v}$, $T(\mathbf{w}) = 2a\mathbf{w} + (\mathbf{v} \times \mathbf{w})$, $T(\mathbf{v} \times \mathbf{w}) = 2a(\mathbf{v} \times \mathbf{w}) + \mathbf{v} \times (\mathbf{v} \times \mathbf{w}) = 2a(\mathbf{v} \times \mathbf{w}) - |\mathbf{v}|^2\mathbf{w}$. Then, if $\mathbf{x} = x\mathbf{v} + y\mathbf{w} + z(\mathbf{v} \times \mathbf{w})$, the given equation is rendered

$$\begin{pmatrix} 2a & 0 & 0 \\ 0 & 2a & -|\mathbf{v}|^2 \\ 0 & 1 & 2a \end{pmatrix} \begin{pmatrix} x \\ y \\ z \end{pmatrix} = -\begin{pmatrix} \lambda \\ 1 \\ 0 \end{pmatrix}.$$

Hence, the components of $\mathbf{x}$ are given by

$$x = -\frac{\lambda}{2a} \qquad y = \frac{-2a}{4a^2 + |\mathbf{v}|^2} \qquad z = \frac{1}{4a^2 + |\mathbf{v}|^2}.$$

Comment. Taking $\mathbf{u} = \lambda\mathbf{v} + \mathbf{w}$, we can rewrite the result in Solution 1 as

$$2a(4a^2 + |\mathbf{v}|^2)\mathbf{x} = \lambda|\mathbf{v}|^2\mathbf{v} + 2a(\mathbf{v} \times \mathbf{w}) - 4a^2\lambda\mathbf{v} - 4a^2\mathbf{w}$$

$$= \lambda(4a^2 + |\mathbf{v}|^2)\mathbf{v} - (2a)^2\mathbf{w} + 2a(\mathbf{v} \times \mathbf{w}),$$

which agrees with the result in Solution 2.

Solution 3. Denote the three components of the vectors by subscripts in the usual way. Since $\mathbf{v} \times \mathbf{x} = (v_2x_3 - v_3x_2, v_3x_1 - v_1x_3, v_1x_2 - v_2x_1)$, the given equation is equivalent to the system of scalar equations:

$$2ax_1 - v_3x_2 + v_2x_3 = -u_1$$
$$v_3x_1 + 2ax_2 - v_1x_3 = -u_2$$
$$-v_2x_1 + v_1x_2 + 2ax_3 = -u_3.$$

The determinant D of the coefficients is equal to $2a(4a^2 + |\mathbf{v}|^2)$, and so is nonzero. By Cramer's Rule, we obtain the solution:

$$Dx_1 = -(u_1v_1 + u_2v_2 + u_3v_3)v_1 + 2a(v_2u_3 - v_3u_2) - 4a^2u_1,$$
$$Dx_2 = -(u_1v_1 + u_2v_2 + u_3v_3)v_2 + 2a(v_3u_1 - v_1u_3) - 4a^2u_2,$$
$$Dx_3 = -(u_1v_1 + u_2v_2 + u_3v_3)v_3 + 2a(v_1u_2 - v_2u_1) - 4a^2u_3.$$

2005:7. Let $f(x)$ be a nonconstant polynomial that takes only integer values when x is an integer, and let P be the set of all primes that divide $f(m)$ for at least one integer m. Prove that P is an infinite set.

The solution to this problem appears in Chap. 11.

2005:8. Let $AX = B$ represent a system of m linear equations in n unknowns, where $A = (a_{ij})$ is an $m \times n$ matrix, $X = (x_1, \ldots, x_n)^t$ is an $n \times 1$ vector and $B = (b_1, \ldots, b_m)^t$ is an $m \times 1$ vector. Suppose that there exists at least one solution for $AX = B$. Given $1 \le j \le n$, prove that the value of the jth component is the same for every solution X of $AX = B$ if and only if the rank of A is decreased if the jth column of A is removed.

Solution 1. Let $C_1, C_2, \ldots, C_n$ be the n columns of A. Suppose that the dimension of the span of these columns is reduced when C_j is removed. Then C_j is not a linear combination of the other C_i. Thus, if $AX = AY = B$ and $Z = X - Y$, then $AZ = O$. This means that $z_1C_1 + z_2C_2 + \cdots + z_jC_j + \cdots + z_nC_n = O$. Because C_j is not a linear combination of the other C_i, we must have that $z_j = 0$, i.e. that $x_j = y_j$.

On the other hand, if the dimension of the span is the same whether or not C_j is present, then C_j must be a linear combination of the remaining C_i. In this case, there is a vector Z with a nonzero value of z_j for which $AZ = z_1C_1 + \cdots + z_jC_j + \cdots + z_nC_n = O$. Thus if X is a solution of $AX = B$, then $X + Z$ is a second solution with a different value of the jth component.

Solution 2. As in the first solution, we observe that every solution of $AX = B$ has the same jth entry if and only if every solution of $AX = O$ has zero in the jth position.

Suppose that every solution of $AX = O$ has 0 in the jth position. Let A_j be the matrix A with the jth column removed and X_j the vector X with the jth entry removed. Then there is a linear isomorphism between the solutions of $AX = 0$ and $A_j X_j = 0$, where $X \to X_j$ is defined by suppressing the jth entry. Since the target space of A_j has dimension $n - 1$. we find that

$$n - \text{rank } A = \text{nullity } A = \text{nullity } A_j = (n - 1) - \text{rank } A_j,$$

whence rank $A_j = \text{rank } A - 1$.

On the other hand, suppose that there are solutions of $AX = O$ for which the jth component fails to vanish. Each solution of $A_j Y = 0$ lifts to a solution of $AX = 0$ by inserting the entry 0 after the $(j - 1)$th position of Y. However, this lifting does not get all the solutions, so that

$$n - \text{rank} A = \text{nullity} A > \text{nullity} A_j = (n - 1) - \text{rank} A_j$$

whence rank$A_j > \text{rank} A - 1$. Since, always, the rank of A_j does not exceed that of A, we must have that the ranks of A_j and A are equal. The desired result follows.

2006:2. Let $\mathbf{u}$ be a unit vector in $\mathbb{R}^3$ and define the operator P by $P(\mathbf{x}) = \mathbf{u} \times \mathbf{x}$ for $\mathbf{x} \in \mathbb{R}^3$ (where $\times$ denotes the cross product).

(a) Describe the operator $I + P^2$.

(b) Describe the action of the operator $I + (\sin \theta)P + (1 - \cos \theta)P^2$.

Solution 1. (a) Recall that $\mathbf{a} \times (\mathbf{b} \times \mathbf{c}) = (\mathbf{a} \cdot \mathbf{c})\mathbf{b} - (\mathbf{a} \cdot \mathbf{b})\mathbf{c}$ for any 3-vectors $\mathbf{a}, \mathbf{b}, \mathbf{c}$.
We have that

$$P^2(\mathbf{x}) = \mathbf{u} \times (\mathbf{u} \times \mathbf{x}) = (\mathbf{u} \cdot \mathbf{x})\mathbf{u} - (\mathbf{u} \cdot \mathbf{u})\mathbf{x} = (\mathbf{u} \cdot \mathbf{x})\mathbf{u} - \mathbf{x}$$

from which we see that $(I + P^2)(\mathbf{x}) = (\mathbf{u} \cdot \mathbf{x})\mathbf{u}$. Thus, $I + P^2$ is an orthogonal projection onto the one-dimensional spaces spanned by $\{\mathbf{u}\}$.

(b) Let $R = I + (\sin \theta)P + (1 - \cos \theta)P^2$. Observe that $P(\mathbf{u}) = \mathbf{O}$. Then $R(\mathbf{u}) = \mathbf{u}$, so that the one-dimensional space spanned by $\{\mathbf{u}\}$ is fixed under R. Let $\mathbf{v}$ be a unit vector orthogonal to $\mathbf{u}$ and let $\mathbf{w} = \mathbf{u} \times \mathbf{v}$. Then $\{\mathbf{u}, \mathbf{v}, \mathbf{w}\}$ is an orthonormal basis for $\mathbb{R}^3$. We have that

$$R(\mathbf{v}) = \mathbf{v} + (\sin \theta)\mathbf{w} + (1 - \cos \theta)(-\mathbf{v}) = (\cos \theta)\mathbf{v} + (\sin \theta)\mathbf{w}$$

and

$$R(\mathbf{w}) = \mathbf{w} + (\sin \theta)(\mathbf{u} \times \mathbf{w}) + (1 - \cos \theta)(-\mathbf{w}) = (-\sin \theta)\mathbf{v} + (\cos \theta)\mathbf{w}.$$

It follows that the plane through $\mathbf{O}$ orthogonal to $\mathbf{u}$ is rotated about the axis $\mathbf{u}$ through an angle θ; by linearity, the same holds for all of $\mathbb{R}^3$.

Solution 2. Wolog, we may assume that $\mathbf{u} = (1, 0, 0)$. Let $\mathbf{x} = (x, y, z)$. Then $\mathbf{u} \times \mathbf{x} = (0, -z, y)$ and $\mathbf{u} \times (\mathbf{u} \times \mathbf{x}) = (0, -y, -z)$. Hence $(I + P^2)\mathbf{x} = (x, 0, 0)$ and

$$(I + (\sin \theta)P + (1 - \cos \theta)P^2)\mathbf{x} = (x, 0, 0) + (0, (\cos \theta)y - (\sin \theta)z, (\sin \theta)y + (\cos \theta)z).$$

Thus, $I + P^2$ is an orthogonal projection onto $\mathbf{u}$ and $I + (\sin \theta)P + (1 - \cos \theta)P^2$ is a rotation about the axis $\mathbf{u}$ through the angle θ.

Solution 3. Let $\mathbf{v}$ be the image of $\mathbf{u}$ after a 90° rotation and let $\mathbf{w} = \mathbf{u} \times \mathbf{v}$. Then $\mathbf{u}, \mathbf{v}, \mathbf{w}$ is an orthonormal basis of $\mathbb{R}^3$ and $\mathbf{u} \times \mathbf{w} = -\mathbf{v}$. Hence, if $\mathbf{x} = x\mathbf{u} + y\mathbf{v} + z\mathbf{w}$, then we find that $(I + P^2)\mathbf{x} = x\mathbf{u} = (\mathbf{x} \cdot \mathbf{u})\mathbf{u}$ and $(I + (\sin\theta)P + (1 - \cos\theta)P^2)(\mathbf{x}) = x\mathbf{u} + [(\cos\theta)y - (\sin\theta)z]\mathbf{v} + [(\cos\theta)z + (\sin\theta)y]\mathbf{w}$, and we reach the same conclusion as in the previous solutions.

2006:7. Let A be a real 3×3 invertible matrix for which the sums of the rows, columns and two diagonals are all equal. Prove that the rows, columns and diagonal sums of A^{-1} are all equal.

Solution 1. Let r be the common sum and let $\mathbf{e} = (1, 1, 1)^{\mathbf{t}}$. Then $A\mathbf{e} = r\mathbf{e}$. Since A is invertible, $r \neq 0$ and $A^{-1}\mathbf{e} = r^{-1}\mathbf{e}$. The other eigenvalues (counting repetitions) of A are u and $-u$, since the trace of A is equal to r. Since A is invertible, $u \neq 0$.

The eigenvalues of A^{-1} are r^{-1}, u^{-1} and $-u^{-1}$, and the trace of A^{-1} is equal to $r^{-1} + u^{-1} - u^{-1} = r^{-1}$. Since also $A^{-1}\mathbf{e} = r^{-1}\mathbf{e}$, the row and main diagonal sums are r^{-1}.

Since $(A^{-1})^{\mathbf{t}} = (A^{\mathbf{t}})^{-1}$, we can apply the same reasoning to $A^{\mathbf{t}}$ to find that the row sums of $(A^{-1})^{\mathbf{t}}$ and, hence, the column sums of A^{-1} are equal to r^{-1}.

Finally, let $B = PA$, where

$$P = \begin{pmatrix} 0 & 0 & 1 \\ 0 & 1 & 0 \\ 1 & 0 & 0 \end{pmatrix}.$$

Then $B^{-1} = A^{-1}P$ and we can apply the result to B, which has the same row, column and diagonal sums as A. The main diagonals of B and B^{-1} are respectively equal to the cross diagonals of A and A^{-1}, so we find that the cross diagonal of A^{-1} is equal to r^{-1}. The desired result follows.

Solution 2. Since the common sum of the rows, columns and diagonals of a magic square is three times the central entry, the general form of the matrix A is

$$\begin{pmatrix} a & 3c - a - b & b \\ c + b - a & c & c + a - b \\ 2c - b & a + b - c & 2c - a \end{pmatrix}$$

The adjugate, adj A of this matrix is

$$\begin{pmatrix} 3c^2 - a^2 + b^2 - ac - 2bc & b^2 - a^2 - 6c^2 + 5ac + bc & 3c^2 + b^2 - a^2 + 2ac - 5bc \\ 5ac - 5bc + b^2 - a^2 & b^2 - a^2 + 2ac - 2bc & b^2 - a^2 - ac + bc \\ b^2 - a^2 - 3c^2 + 2ac + bc & 6c^2 + b^2 - a^2 - ac - 5bc & b^2 - a^2 - 3c^2 + 5ac - 2bc \end{pmatrix}.$$

The sum of each row, column and diagonal of adj A is $3(b - a)(b + a - 2c)$. Since $(\det A)A^{-1} = \text{adj } A$, the desired result follows.

Comment. A direct solution for the rows and columns can be obtained from the relationships between the elements of A and A^{-1}. Let $A = (a_{ij})$ and $A^{-1} = (b_{ij})$. Then, for $1 \leq i, k \leq 3$,

$$\sum_{j=1}^{3} a_{ij}b_{jk} = \delta_{ik},$$

where δ_{ij} is the Kronecker delta. Let the common sums for A be s. Then

$$s\sum_{j=1}^{3} b_{jk} = \sum_{j=1}^{3}(\sum_{i=1}^{3} a_{ij})b_{jk} = \sum_{i=1}^{3}(\sum_{j=1}^{3} a_{ij}b_{jk}) = \sum_{i=1}^{3} \delta_{ik} = 1,$$

whence $s \neq 0$ and $\sum_{j=1}^{3} b_{jk} = 1/s$. Hence, each column sum of A^{-1} is $1/s$. A similar argument holds for the row sums.

2007:7. Find the Jordan canonical form of the matrix $\mathbf{uv}^t$ where $\mathbf{u}$ and $\mathbf{v}$ are column vectors in $\mathbb{C}^n$.

Solution. Suppose first that $\mathbf{v}^t\mathbf{u} \neq 0$. Then

$$(\mathbf{uv}^t)\mathbf{u} = \mathbf{u}(\mathbf{v}^t\mathbf{u}) = (\mathbf{v}^t\mathbf{u})\mathbf{u}$$

with $\mathbf{u} \neq \mathbf{O}$, so that $\mathbf{uv}^t$ has the nonzero eigenvalue $\mathbf{v}^t\mathbf{u}$. Since $\mathbf{uv}^t$ has rank 1, so also does its Jordan form, which is then

$$\begin{pmatrix} \mathbf{v}^t\mathbf{u} & 0 & 0 & \cdots & 0 \\ 0 & 0 & 0 & \cdots & 0 \\ \vdots & \vdots & \vdots & \ddots & \vdots \\ 0 & 0 & 0 & \cdots & 0 \end{pmatrix}$$

If $\mathbf{v}^t\mathbf{u} = 0$, $\mathbf{u} \neq \mathbf{O}$, $\mathbf{v} \neq \mathbf{O}$, then $(\mathbf{uv}^t)^2 = \mathbf{O}$ and $\mathbf{uv}^t$ has rank 1, so that its Jordan form is

$$\begin{pmatrix} 0 & 1 & 0 & \cdots & 0 \\ 0 & 0 & 0 & \cdots & 0 \\ \vdots & \vdots & \vdots & \ddots & \vdots \\ 0 & 0 & 0 & \cdots & 0 \end{pmatrix}.$$

2008:5. Suppose that $a, b, c \in \mathbb{C}$ with $ab = 1$. Evaluate the determinant of

$$\begin{pmatrix} c & a & a^2 & \cdots & a^{n-1} \\ b & c & a & \cdots & a^{n-2} \\ b^2 & b & c & \cdots & a^{n-3} \\ \vdots & \vdots & \vdots & & \vdots \\ b^{n-1} & b^{n-2} & & \cdots & c \end{pmatrix}.$$

Solution 1. With $\mathbf{u} = (1, b, b^2, \cdots, b^{n-1})$ and $\mathbf{v} = (1, a, a^2, \cdots, a^{n-1})$, we have to find $\det(\mathbf{u}^t\mathbf{v} + (c-1)I_n)$ with $\mathbf{u}\mathbf{v}^t = n$. Since $\mathbf{u}^t\mathbf{v}$ is similar to

$$\begin{pmatrix} n & 0 & 0 & \cdots & 0 \\ 0 & 0 & 0 & \cdots & 0 \\ \vdots & \vdots & \vdots & & \vdots \\ 0 & 0 & 0 & \cdots & 0 \end{pmatrix}$$

we are led to the answer $(n + c - 1)(c - 1)^{n-1}$.

Solution 2. Setting $c = 1$ yields n columns that are proportional. Hence $(c-1)^{n-1}$ is a factor of the determinant. The leading term is c^n, so that the determinant must be $(c+k)(c-1)^{n-1}$. Since the coefficient of c^{n-1} vanishes, $k = n - 1$.

Solution 3. [C. Ochanine] Denote the determinant of the $n \times n$ matrix by $\Delta_n(a, b, c)$. By subtracting a suitable multiple of the first row, we find that $\Delta_n(a, b, c)$ is equal to the determinant of

$$\begin{pmatrix} c & a & a^2 & \cdots & a^{n-1} \\ 0 & c - \frac{1}{c} & a - \frac{a}{c} & \cdots & a^{n-2} - \frac{a^{n-2}}{c} \\ 0 & b - \frac{b}{c} & c - \frac{1}{c} & \cdots & a^{n-3} - \frac{a^{n-3}}{c} \\ 0 & \vdots & \vdots & & \vdots \end{pmatrix}$$

Noting that $c - \frac{1}{c} = (1 - \frac{1}{c})(c + 1)$ and pulling out the factor $1 - \frac{1}{c}$ from the last $n - 1$ rows yields

$$\Delta_n(a, b, c) = c\left(1 - \frac{1}{c}\right)^{n-1} \Delta_{n-1}(a, b, c + 1).$$

An induction argument shows that $\Delta_n(a, b, c) = (n + c - 1)(c - 1)^{n-1}$.

Solution 4. [S. Wong] The eigenvalues of the given matrix are those numbers λ for which

$$0 = \det \begin{pmatrix} c - \lambda & a & a^2 & \cdots & a^{n-1} \\ b & c - \lambda & a & \cdots & a^{n-2} \\ b^2 & b & c - \lambda & \cdots & a^{n-3} \\ \vdots & \vdots & \vdots & & \vdots \\ b^{n-1} & b^{n-2} & & \cdots & c - \lambda \end{pmatrix}$$

$$= \det \begin{pmatrix} 1 - (\lambda - c + 1) & a & a^2 & \cdots & a^{n-1} \\ b & 1 - (\lambda - c + 1) & a & \cdots & a^{n-2} \\ b^2 & b & 1 - (\lambda - c + 1) & \cdots & a^{n-3} \\ \vdots & \vdots & \vdots & & \vdots \\ b^{n-1} & b^{n-2} & & \cdots & 1 - (\lambda - c + 1) \end{pmatrix}.$$

Hence λ is an eigenvalue of the given matrix if and only if $\lambda - c + 1$ is an eigenvalue with the same multiplicity of the matrix

$$M \equiv \begin{pmatrix} 1 & a & a^2 & \cdots & a^{n-1} \\ b & 1 & a & \cdots & a^{n-2} \\ b^2 & b & 1 & \cdots & a^{n-3} \\ \vdots & \vdots & \vdots & & \vdots \\ b^{n-1} & b^{n-2} & & \cdots & 1 \end{pmatrix}.$$

Since all the columns of M are in proportion, the rank of M is 1 and so 0 is an eigenvalue of multiplicity $n - 1$. By examining the trace (which is the sum of the eigenvalues), we see that the remaining eigenvalue is n (an eigenvector is $(1, b, b^2, \cdots, b^{n-1})^t$). Thus, the eigenvalues of the given matrix are $c - 1$ with multiplicity $n - 1$ and $n + c - 1$ with multiplicity 1. Since the determinant of the given matrix is the product of its eigenvalues, we find that the required answer is $(n + c - 1)(c - 1)^{n-1}$.

Solution 5. [E. Flat] Denote the determinant to be found by D_n. Observe that $D_1 = c$ and that $D_2 = c^2 - 1 = (c - 1)(c + 1)$. By taking b times each row from the next, we find that

$$D_n = \det \begin{pmatrix} c & a & a^2 & \cdots & a^{n-2} & a^{n-1} \\ b(1-c) & c-1 & 0 & \cdots & 0 & 0 \\ 0 & b(1-c) & c-1 & \cdots & 0 & 0 \\ \vdots & \vdots & \vdots & \cdots & \vdots & \vdots \\ 0 & 0 & 0 & \cdots & c-1 & 0 \\ 0 & 0 & 0 & \cdots & b(1-c) & c-1 \end{pmatrix}.$$

Expanding along the last column yields, for $n \geq 2$,

$$D_n = (c - 1)D_{n-1} + (-1)^{n-1}a^{n-1}[b(1 - c)]^{n-1} = (c - 1)D_{n-1} + (c - 1)^{n-1}.$$

An induction argument then establishes that

$$D_n = (c - 1)^{n-1}(c + n - 1).$$

Solution 6. [J. Kramar; O. Ivrii] The (i, j)th term in the given matrix is $c^{\delta_{ij}}a^{j-i}$, where δ_{ij} is the Kronecker delta that takes the value 1 when $i = j$ and 0 otherwise. The expansion of the determinant of this matrix is the sum of $n!$ terms of the form

$$\pm c^{\epsilon(\pi)} \prod_{i=1}^n a^{i-\pi(i)} = \pm c^{\epsilon(\pi)} \prod_{i=1}^n a^i \prod_{i=1}^n a^{-\pi(i)} = \pm c^{\epsilon(\pi)}.$$

where π is a permutation of $\{1, 2, \ldots, n\}$ and $\epsilon(\pi)$ is the number of fixed elements of π. Thus, the value of the determinant is independent of a and b, so that, wolog, we may restrict ourselves to the case that $a = b = 1$.

But then the required determinant is the product of the eigenvalues (counting multiplicity) of

$$
\begin{pmatrix}
c & 1 & 1 & \cdots & 1 \\
1 & c & 1 & \cdots & 1 \\
1 & 1 & c & \cdots & 1 \\
\vdots & \vdots & \vdots & & \vdots \\
1 & 1 & 1 & \cdots & c
\end{pmatrix}.
$$

The $(n-1)$-dimensional subspace

$$\{(x_1, x_2, \ldots, x_n)^t : x_1 + x_2 + \cdots + x_n = 0\}$$

consists of eigenvectors with eigenvalue $c-1$. A complement of this subspace is the span of the vector $(1, 1, \ldots, 1)^t$, which is an eigenvector with eigenvalue $c + n - 1$. Therefore the product of the eigenvalues and determinant of the matrix is $(c + n - 1)(c - 1)^{n-1}$.

2009:2. Let n and k be integers with $n \geq 0$ and $k \geq 1$. Let $\mathbf{x}_0$, $\mathbf{x}_1$, $\ldots$, $\mathbf{x}_n$ be $n+1$ distinct points in $\mathbb{R}^k$ and let y_0, y_1, $\ldots$, y_n be $n+1$ real numbers (not necessarily distinct). Prove that there exists a polynomial p of degree at most n in the coordinates of $\mathbf{x}$ with respect to the standard basis for which $p(\mathbf{x}_i) = y_i$ for $0 \leq i \leq n$.

Solution. For $0 \leq i < j \leq n$, let H_{ij} be the hyperplane $\{\mathbf{z} : (\mathbf{x}_i - \mathbf{x}_j) \cdot \mathbf{z} = 0\}$. Since there are finitely many such hyperplanes, their union is not all of $\mathbf{R}^k$. (See Problem **2002:7**.) Therefore, there exists a vector $\mathbf{u}$ for which $(\mathbf{x}_i - \mathbf{x}_j) \cdot \mathbf{u} \neq 0$ for all distinct i, j. Therefore, the real numbers $t_i = \mathbf{x_i} \cdot \mathbf{u}$ are all distinct $0 \leq i \leq n$. There is a real polynomial q of degree at most n for which $q(t_i) = y_i$ $(0 \leq i \leq n)$. The polynomial $p(\mathbf{x}) = q(\mathbf{x} \cdot \mathbf{u})$ has the desired property.

2009:5. Find a 3×3 matrix A with elements in $\mathbb{Z}_2$ for which $A^7 = I$ and $A \neq I$. (Here, I is the identity matrix and $\mathbb{Z}_2$ is the field of two elements 0 and 1 where addition and multiplication are defined modulo 2.)

Solution. The minimum polynomial of A has degree at most 3 and divides the polynomial

$$t^7 - 1 = (t-1)(t^6 + t^5 + t^4 + t^3 + t^2 + t + 1).$$

Since A is not the identity matrix, its minimum polynomial divides the latter factor. This minimum polynomial cannot be any of the irreducibles t, $t+1$ (by the Factor Theorem) and $t^2 + t + 1$, so it must be one of the irreducibles $t^3 + t^2 + 1$, $t^3 + t + 1$. Indeed

$$t^6 + t^5 + t^4 + t^3 + t^2 + t + 1 = (t^3 + t + 1)(t^3 + t^2 + 1),$$

over $\mathbb{Z}_2$.

We let A be the companion matrix of the first factor on the right, namely

$$A = \begin{pmatrix} 0 & 1 & 0 \\ 0 & 0 & 1 \\ 1 & 1 & 0 \end{pmatrix}.$$

Indeed, it can be checked that $A^3 = I + A$, whence $A^6 = I + A^2$ and $A^7 = A + A^3 = A + I + A = I$, as desired.

[Problem **2009:5** was contributed by Alfonso Grazia-Saz.]

2009:8. Let a, b, c be members of a real inner-product space $(V, \langle, \rangle)$ whose norm is given by $\|x\|^2 = \langle x, x \rangle$. (You may assume that V is $\mathbb{R}^n$ if you wish.) Prove that

$$\|a + b\| + \|b + c\| + \|c + a\| \le \|a\| + \|b\| + \|c\| + \|a + b + c\|$$

for $a, b, c \in V$.

Solution. From the inner product representations of the squares of the norms, we have that

$$\|a + b\|^2 + \|b + c\|^2 + \|c + a\|^2 = \|a\|^2 + \|b\|^2 + \|c\|^2 + \|a + b + c\|^2.$$

Squaring both sides of the desired inequality, we see that it suffices to prove the following

$$(7.1) \qquad \|p + r\|\|q + r\| \le \|p\|\|q\| + \|r\|\|p + q + r\|$$

for $p, q, r \in V$ and apply this to permutations of a, b, c.

Making the substitution $x = p + r$, $y = q + r$, $z = p + q + r$, so that $p = z - y$, $q = z - x$, $r = x + y - z$, we see that (7.1) is equivalent to

$$(7.2) \qquad \|x\|\|y\| \le \|z - x\|\|z - y\| + \|z\|\|x + y - z\|$$

for $x, y, z, \in V$.

Let w be the orthogonal projection of z onto the span of x and y. Then $z = w + v$ where v is orthogonal to x, y and w, so that

$$\|z - x\|^2 = \|(w - x) + v\|^2 = \|w - x\|^2 + \|v\|^2 \ge \|w - x\|^2$$

and $\|z - x\| \ge \|w - x\|$. Similarly, $\|z - y\| \ge \|w - y\|$, $\|z\| \ge \|w\|$ and $\|x + y - z\| \ge \|x + y - w\|$. Thus, it is enough to prove that

$$\|x\|\|y\| \le \|w - x\|\|w - y\| + \|w\|\|x + y - w\|$$

where x, y, w belong to a two-dimensional real inner product space.

But such a space is isometric to $\mathbb{C}$ with the usual absolute value, so we may suppose that $x, y, w \in \mathbf{C}$ and can be multiplied. Since

$$xy = (w - x)(w - y) + w(x + y - w),$$

an application of the triangle inequality yields the result.

Comment. Geometrically, equation (7.2) can be formulated as: *suppose that $OABC$ is a parallelogram and that P is a point in space; then* $|OA||OC| \leq |PA||PC| + |PO||PB|$.

2010:3. Let $\mathbf{a}$ and $\mathbf{b}$, the latter nonzero, be vectors in $\mathbb{R}^3$. Determine the value of λ for which the vector equation

$$\mathbf{a} - (\mathbf{x} \times \mathbf{b}) = \lambda \mathbf{b}$$

is solvable, and then solve it.

Solution 1. If there is a solution, we must have $\mathbf{a} \cdot \mathbf{b} = \lambda |\mathbf{b}|^2$, so that $\lambda = (\mathbf{a} \cdot \mathbf{b})/|\mathbf{b}|^2$. On the other hand, suppose that λ has this value. Then

$$\mathbf{0} = \mathbf{b} \times \mathbf{a} - \mathbf{b} \times (\mathbf{x} \times \mathbf{b})$$
$$= \mathbf{b} \times \mathbf{a} - [(\mathbf{b} \cdot \mathbf{b})\mathbf{x} - (\mathbf{b} \cdot \mathbf{x})\mathbf{b}]$$

so that

$$\mathbf{b} \times \mathbf{a} = |\mathbf{b}|^2 \mathbf{x} - (\mathbf{b} \cdot \mathbf{x})\mathbf{b}.$$

A particular solution of this equation is

$$\mathbf{x} = \mathbf{u} \equiv \frac{\mathbf{b} \times \mathbf{a}}{|\mathbf{b}|^2}.$$

Let $\mathbf{x} = \mathbf{z}$ be any other solution. Then

$$|\mathbf{b}|^2(\mathbf{z} - \mathbf{u}) = |\mathbf{b}|^2\mathbf{z} - |\mathbf{b}|^2\mathbf{u}$$
$$= (\mathbf{b} \times \mathbf{a} + (\mathbf{b} \cdot \mathbf{z})\mathbf{b}) - (\mathbf{b} \times \mathbf{a} + (\mathbf{b} \cdot \mathbf{u})\mathbf{b})$$
$$= (\mathbf{b} \cdot \mathbf{z})\mathbf{b}$$

so that $\mathbf{z} - \mathbf{u} = \mu\mathbf{b}$ for some scalar μ.

We check when this works. Let $\mathbf{x} = \mathbf{u} + \mu\mathbf{b}$ for some scalar μ. Then

$$\mathbf{a} - (\mathbf{x} \times \mathbf{b}) = \mathbf{a} - (\mathbf{u} \times \mathbf{b}) = \mathbf{a} - \frac{(\mathbf{b} \times \mathbf{a}) \times \mathbf{b}}{|\mathbf{b}|^2}$$
$$= \mathbf{a} + \frac{\mathbf{b} \times (\mathbf{b} \times \mathbf{a})}{|\mathbf{b}|^2}$$
$$= \mathbf{a} + \frac{(\mathbf{b} \cdot \mathbf{a})\mathbf{b} - (\mathbf{b} \cdot \mathbf{b})\mathbf{a}}{|\mathbf{b}|^2}$$
$$= \mathbf{a} + \left(\frac{\mathbf{b} \cdot \mathbf{a}}{|\mathbf{b}|^2}\right)\mathbf{b} - \mathbf{a} = \lambda\mathbf{b},$$

as desired. Hence, the solution is

$$\mathbf{x} = \frac{\mathbf{b} \times \mathbf{a}}{|\mathbf{b}|^2} + \mu\mathbf{b},$$

where μ is an arbitrary scalar.

Solution 2. [B. Yahagni] Suppose, to begin with, that $\{\mathbf{a}, \mathbf{b}\}$ is linearly dependent. Then $\mathbf{a} = [(\mathbf{a} \cdot \mathbf{b})/|\mathbf{b}|^2]\mathbf{b}$. Since $(\mathbf{x} \times \mathbf{b}) \cdot \mathbf{b} = 0$ for all $\mathbf{x}$, the equation has no solutions except when $\lambda = (\mathbf{a} \cdot \mathbf{b})/|\mathbf{b}|^2$. In this case, it becomes $\mathbf{x} \times \mathbf{b} = \mathbf{0}$ and is satisfied by $\mathbf{x} = \mu\mathbf{b}$, where μ is any scalar.

Otherwise, $\{\mathbf{a}, \mathbf{b}, \mathbf{a} \times \mathbf{b}\}$ is linearly independent and constitutes a basis for $\mathbb{R}^3$. Let a solution be

$$\mathbf{x} = \alpha\mathbf{a} + \mu\mathbf{b} + \beta(\mathbf{a} \times \mathbf{b}).$$

Then

$$\mathbf{x} \times \mathbf{b} = \alpha(\mathbf{a} \times \mathbf{b}) + \beta[(\mathbf{a} \times \mathbf{b}) \times \mathbf{b}] = \alpha(\mathbf{a} \times \mathbf{b}) + \beta(\mathbf{a} \cdot \mathbf{b})\mathbf{b} - \beta(\mathbf{b} \cdot \mathbf{b})\mathbf{a}$$

and the equation becomes

$$(1 + \beta|\mathbf{b}|^2)\mathbf{a} - \beta(\mathbf{a} \cdot \mathbf{b})\mathbf{b} - \alpha(\mathbf{a} \times \mathbf{b}) = \lambda\mathbf{b}.$$

Therefore $\alpha = 0$, μ is arbitrary, $\beta = -1/|\mathbf{b}|^2$ and $\lambda = -\beta(\mathbf{a} \cdot \mathbf{b}) = (\mathbf{a} \cdot \mathbf{b})/|\mathbf{b}|^2$.

Therefore, the existence of a solution requires that $\lambda = (\mathbf{a} \cdot \mathbf{b})/|\mathbf{b}|^2$ and the solution then is

$$\mathbf{x} = \mu\mathbf{b} - \frac{1}{|\mathbf{b}|^2}(\mathbf{a} \times \mathbf{b}).$$

Solution 3. Writing the equation in vector components yields the system

$$b_3 x_2 - b_2 x_3 = a_1 - \lambda b_1;$$
$$-b_3 x_1 + b_1 x_3 = a_2 - \lambda b_2;$$
$$b_2 x_1 - b_1 x_2 = a_3 - \lambda b_3.$$

The matrix of coefficients of the left side is of rank 2, so that the corresponding homogeneous system of equations has a single infinity of solutions. Multiplying the three equations by b_1, b_2 and b_3 respectively and adding yields

$$0 = a_1 b_1 + a_2 b_2 + a_3 b_3 - \lambda(b_1^2 + b_2^2 + b_3^2).$$

Thus, for a solution to exist, we require that

$$\lambda = \frac{a_1 b_1 + a_2 b_2 + a_3 b_3}{b_1^2 + b_2^2 + b_3^2}.$$

In addition, we learn that the corresponding homogeneous system is satisfied by

$$(x_1, x_2, x_3) = \mu(b_1, b_2, b_3)$$

where μ is an arbitrary scalar.

It remains to find a particular solution for the nonhomogeneous system. Multiplying the third equation by b_2 and subtracting the second multiplied by b_3, we obtain that

$$(b_2^2 + b_3^2)x_1 = b_1(b_2 x_2 + b_3 x_3) + (a_3 b_2 - a_2 b_3).$$

Therefore, setting $b_1^2 + b_2^2 + b_3^2 = b^2$, we have that

$$b^2 x_1 = b_1(b_1 x_1 + b_2 x_2 + b_3 x_3) + (a_3 b_2 - a_2 b_3).$$

Similarly

$$b^2 x_2 = b_2(b_1 x_1 + b_2 x_2 + b_3 x_3) + (a_1 b_3 - a_3 b_1),$$
$$b^2 x_3 = b_3(b_1 x_1 + b_2 x_2 + b_3 x_3) + (a_2 b_1 - a_1 b_2).$$

Observing that $b_1 x_1 + b_2 x_2 + b_3 x_3$ vanishes when

$$(x_1, x_2, x_3) = (a_3 b_2 - a_2 b_3, a_1 b_3 - a_3 b_1, a_2 b_1 - a_1 b_2),$$

we obtain a particular solution to the system:

$$(x_1, x_2, x_3) = b^{-2}(a_3b_2 - a_2b_3, a_1b_3 - a_3b_1, a_2b_1 - a_1b_2).$$

Adding to this the general solution of the homogeneous system yields the solution of the nonhomogeneous system.

2010:8. Let A be an invertible symmetric $n \times n$ matrix with entries $\{a_{ij}\}$ in $\mathbb{Z}_2$. Prove that there is an $n \times n$ matrix M with entries in $\mathbb{Z}_2$ such that $A = M^t M$ only if $a_{i,i} \neq 0$ for some i.

Solution 1. [A. Kim] Let the entries of M be $m_{i,j}$. Then

$$a_{ii} = \sum_j m_{ji}m_{ji} = \sum_j m_{ji}^2 = \sum_j m_{ji}.$$

Suppose that $a_{ii} = 0$ for each i. Then each column must have evenly many entries equal to 1. But then the sum of the row vectors of M must be the zero vector, and so the rows are linearly dependent. Hence the rank of M is less than n, and so M is not invertible.

Solution 2. Let $x = (x_1, x_2, \ldots, x_n)^t$ be a column vector over $\mathbb{Z}_2$ and observe that

$$xAx^t = \sum_{i=1}^n a_{ii}x_i^2 = (\sum_{i=1}^n a_{ii}x_i)^2.$$

(Note that $a_{ij} + a_{ji} = 2a_{ij} = 0$.) Define $N = \{x : xAx^t = 0\}$. Then $N = \{x : \sum_{i=1}^n a_{ii}x_i = 0\}$.

Suppose that $A = M^t M$. Then M is invertible and so $x \to Mx$ is a surjection (onto). Therefore the equation

$$0 = xAx^t = xM^t Mx^t = (Mx^t)^t(Mx^t)$$

is not satisfied for each x, so that N is a proper subspace of $(\mathbb{Z}_2)^n$. Therefore, there must exist i for which $a_{i,i} \neq 0$.

[Problem **2010:8** was contributed by Franklin Vera Pachebo.]

2011:7. Suppose that there are 2011 students in a school and that each student has a certain number of friends among his schoolmates. It is assumed that if A is a friend of B, then B is a friend of A, and also that there may exist certain pairs that are not friends. Prove that there is a nonvoid subset S of these students for which every student in the school has an even number of friends in S.

Solution. Define a 2011×2011 matrix with entries a_{ij} in $\mathbb{Z}_2 \equiv \{0, 1\}$, the field of integers modulo 2, as follows: $a_{ij} = 1$ if and only if i and j are distinct and persons i and j are friends, and $a_{ij} = 0$ otherwise.

The determinant of this matrix is equal to

$$\sum a_{1,\sigma(1)}a_{2,\sigma(2)} \cdots a_{2011,\sigma(2011)},$$

where the sum has $n!$ terms and is taken over all permutations σ of the symmetric group S_{2011}. Since 2011 is odd, any permutation of period 2, consisting of a product of independent transpositions, must leave at least one element fixed so that its contribution to the determinant sum is 0. The remaining permutations come in inverse pairs that yield the same product value in the sum. It follows that the determinant of the matrix has the value 0.

Therefore there is a linearly dependent subset of rows, and therefore a set of rows whose modulo 2 sum is 0. The portion of each column belonging to each of these rows must contain an even number of 1's. This means that the person corresponding to that column must have evenly many friends among the persons corresponding to the rows.

[Problem **2011:7** was contributed by Ali Feiz Mohammadi.]

2011:9. Suppose that A and B are two square matrices of the same order for which the indicated inverses exist. Prove that

$$(A + AB^{-1}A)^{-1} + (A + B)^{-1} = A^{-1}.$$

Solution 1. [V. Baydina]

$$
\begin{aligned}
(A + AB^{-1}A)^{-1} + (A + B)^{-1} &= [A(I + B^{-1}A)]^{-1} + (A + B)^{-1} \\
&= (B^{-1}B + B^{-1}A)^{-1}A^{-1} + (B + A)^{-1} \\
&= [B^{-1}(B + A)]^{-1}A^{-1} + (B + A)^{-1} \\
&= (B + A)^{-1}(BA^{-1} + I) \\
&= (B + A)^{-1}(BA^{-1} + AA^{-1}) \\
&= (B + A)^{-1}(B + A)A^{-1} = A^{-1}.
\end{aligned}
$$

Solution 2. [K. Ng]

$$
\begin{aligned}
[A + AB^{-1}A][A^{-1} - (A + B)^{-1}] &= I - A(A + B)^{-1} + AB^{-1} - AB^{-1}A(A + B)^{-1} \\
&= I + AB^{-1} - A(I + B^{-1}A)(A + B)^{-1} \\
&= I + AB^{-1} - A(I + B^{-1}A)[B(I + B^{-1}A)]^{-1} \\
&= I + AB^{-1} - A(I + B^{-1}A)(I + B^{-1}A)^{-1}B^{-1} \\
&= I + AB^{-1} - AB^{-1} = I.
\end{aligned}
$$

Hence $A^{-1} - (A + B)^{-1} = (A + AB^{-1}A)^{-1}$ as desired.

Solution 3. Fix the matrix A and define a new multiplication of the set of square matrices by $X * Y = XA^{-1}Y$. This multiplication is associative and distributive with respect to ordinary matrix addition. For any matrix X, we have that $X * A = X = A * X$ so that A is the new identity matrix; the inverse of a nonsingular matrix X is equal to $X^{\circ} \equiv AX^{-1}A$.

Observe that

$$(A + B) * (A + B^\circ) = (A + B)A^{-1}(A + B^\circ) = (I + BA^{-1})(A + B^\circ)$$
$$= (A + B) + (B^\circ + BA^{-1}B^\circ) = (A + B) + (B^\circ + A)$$
$$= (A + B) + (A + B^\circ).$$

$*$—multiplying this equality on the left by $(A + B)^\circ$ and on the right by $(A + B^\circ)^\circ$ yields that

$$A = (A + B^\circ)^\circ + (A + B)^\circ = A(A + AB^{-1}A)^{-1}A + A(A + B)^{-1}A.$$

Multiplying this equation normally on the left and on the right by A^{-1} yields the desired result.

2012:10. Let A be a square matrix whose entries are complex numbers. Prove that $A^* = A$ if and only if $AA^* = A^2$.

Comment. If $A = A^*$, then clearly $AA^* = A^2$. Henceforth, we assume that $AA^* = A^2$ and that A is a $n \times n$ matrix.

Solution 1. Let $B = i(A - A^*)$, so that B is hermitian and $AB = O$. Then B is unitarily equivalent to a diagonal matrix, so that it suffices to show that all the eigenvalues of B are 0.

Suppose, if possible, that B has a nonzero eigenvalue λ with nontrivial eigenvector x. Then $O = ABx = \lambda Ax$, so that

$$O = Ax = A^*x - i\lambda x.$$

Therefore

$$0 = x^* Ax = x^* A^* x - i\lambda x^* x.$$

Since

$$x^* A^* x = \langle x, A^* x \rangle = \overline{\langle A^* x, x \rangle} = \overline{\langle x, Ax \rangle} = \overline{x^* Ax} = 0,$$

we have that $i\lambda \|x\|^2 = i\lambda x^* x = 0$, which yields a contradiction. Therefore $B = O$ and $A = A^*$.

Solution 2. There exists a unitary matrix U (with $UU^* = U^*U = I$) for which $A = U^*TU$ for some upper triangular matrix T. Let the diagonal of T be $(\lambda_1, \lambda_2, \ldots, \lambda_n)$ and its other entries be t_{ij} for $1 \leq i < j \leq n$.

We have that

$$TT^* = UAU^*UA^*U^* = UAA^*U^* = UA^2U^* = UAU^*UAU^* = T^2,$$

so that the trace of TT^* equals the trace of T^2. Thus,

$$\sum_{k=1}^n |\lambda_k|^2 + \sum_{1 \leq i < j \leq n} |t_{ij}|^2 = \sum_{k=1}^n \lambda_k^2.$$

The right side of this equation must be real, and

$$\sum_{k=1}^n \lambda_k^2 \leq |\sum_{k=1}^n \lambda_k|^2 \leq \sum_{k=1}^n |\lambda_k|^2,$$

so that each t_{ij} is zero and $\lambda_k^2 = |\lambda_k|^2$ for each k. Thus T is a diagonal matrix with real eigenvalues, and so hermitian. Therefore A is hermitian.

Solution 3. Observe that

$$A^{*2} = (A^2)^* = (AA^*)^* = AA^* = A^2.$$

Suppose that it has been established that $A^{*k} = A^k$ for some integer k not less than 2. Then

$$A^{k+1} = A(A^{*k}) = (AA^*)(A^{*k-1}) = A^{*2}(A^{*k-1}) = A^{*k+1}$$

so that $A^{*m} = A^m$ for all integers $m \geq 2$.

Let p be the minimal polynomial for A. Since every eigenvalue of $A^2 = AA^*$ is nonnegative, it follows that every eigenvalue of A is real, so that p has real coefficients and $p(A^*) = (p(A))^* = O$. Thus p is the minimal polynomial for A^*.

Suppose, if possible, that $p(t) = t^2 q(t)$, for some polynomial $q(t)$. Let x be any vector, and let $y = q(A^*)x$. Then

$$AA^*y = A^2y = A^{*2}y = p(A^*)x = 0$$

so that

$$\langle A^*y, A^*y \rangle = \langle y, AA^*y \rangle = 0$$

from which $A^*q(A^*)x = A^*y = 0$. But this contradicts the fact that p is the minimal polynomial for A^*. Thus $p(t) = a_0 + a_1 t + a_2 t^2 + \cdots$ where at least one of a_0 and a_1 is nonzero.

Suppose that $a_0 = 0$. Then

$$O = p(A) - p(A^*) = a_1(A - A^*)$$

so that $A = A^*$. On the other hand, if a_0 is nonzero, then

$$O = Ap(A) - A^*p(A^*) = a_0(A - A^*) + a_1 A^2 - a_1 A^{*2} = a_0(A - A^*)$$

and again $A = A^*$. Thus, the desired result holds.

Solution 4. Suppose that $M = (m_{ij})$ is an arbitrary $n \times n$ matrix. Then $\text{tr}(M)$, the trace of M (the sum of its diagonal elements), is equal to the complex conjugate of $\text{tr}(M^*)$. Since the ith diagonal element of MM^* is equal to $\sum_j |m_{ij}|^2$, it follows that $\text{tr}(MM^*) = \sum_{i,j} |m_{ij}|^2$ and we have that $\text{tr}(MM^*) = \text{tr}(M^*M)$.

It follows from this that $M = O$ if and only if $\text{tr}(MM^*) = 0$. If $M = M^*$, then its trace is real and $\text{tr}(M) = \text{tr}(M^*)$.

Consider the situation of the problem where $AA^* = A^2$. We have that

$$(A - A^*)(A - A^*)^* = (A - A^*)(A^* - A) = (AA^* - A^2) + (A^*A - A^{*2}) = A^*A - A^{2*},$$

from which we find that

$$\text{tr}((A - A^*)(A - A^*)^*) = \text{tr}(A^*A) - \text{tr}(A^{2*}) = \text{tr}(AA^*) - \overline{\text{tr}(A^2)}$$

$$= \text{tr}(AA^*) - \overline{\text{tr}(AA^*)} = 0$$

since the trace of AA^* is real. It follows that $A - A^* = O$ so that $A = A^*$.

[Problem **2012:10** and its solutions are drawn from the *American Mathematical Monthly* 101:4 (April, 1994), 362; 104:3 (1997), 277–278.]

2013:7. Let $(V, \langle \cdot \rangle)$ be a two-dimensional inner product space over the complex field $\mathbb{C}$ and let z_1 and z_2 be unit vectors in V. Prove that

$$\sup\{|\langle z, z_1 \rangle \langle z, z_2 \rangle| : \|z\| = 1\} \geq \frac{1}{2}$$

with equality if and only if $\langle z_1, z_2 \rangle = 0$.

Solution. Let $\langle z_1, z_2 \rangle = a + bi$ for real a and b. Since $|\langle z, z_1 \rangle \langle z, z_2 \rangle| = |\langle z, z_1 \rangle \langle z, -z_2 \rangle|$, there is no loss of generality in assuming that $a = \text{Re}\langle z_1, z_2 \rangle \geq 0$. Suppose that $w = (z_1 + z_2)/(\|z_1 + z_2\|)$. Then $\|w\| = 1$ and the supremum in the problem is not less than $\|\langle w, z_1 \rangle \langle w, z_2 \rangle\|$. Note that $|\langle w, z_1 \rangle| = |1 + \langle z_2, z_1 \rangle|/(\|z_1 + z_2\|)$ and that $|\langle w, z_2 \rangle| = |1 + \langle z_1, z_2 \rangle|/(\|z_1 + z_2\|)$. Also $\|z_1 + z_2\| = \langle z_1 + z_2, z_1 + z_2 \rangle = \langle z_1, z_1 \rangle + \langle z_2, z_2 \rangle + \langle z_1, z_2 \rangle + \langle z_2, z_1 \rangle = 2 + 2\text{Re}\langle z_1, z_2 \rangle = 2 + 2a$.

$$
\begin{aligned}
|\langle w, z_1 \rangle \langle w, z_2 \rangle| &= \frac{|1 + \langle z_1, z_2 \rangle|^2}{2 + 2a} \\
&= \frac{(1+a)^2 + b^2}{2(1+a)} \\
&= \frac{1}{2}\left(1 + a + \frac{b^2}{1+a}\right) \geq \frac{1}{2},
\end{aligned}
$$

with equality if and only if $a = b = 0$, i.e. $\langle z_1, z_2 \rangle = 0$. Thus, if the supremum is $\frac{1}{2}$, then $\langle z_1, z_2 \rangle = 0$.

On the other hand, suppose that $\langle z_1, z_2 \rangle = 0$. Then $\{z_1, z_2\}$ is an orthonormal basis, and we can write $z = uz_1 + vz_2$ where $|u|^2 + |v|^2 = 1$ for $\|z\| = 1$. Then

$$|\langle z, z_1 \rangle \langle z, z_2 \rangle| = |u||v| \leq \frac{1}{2}(|u|^2 + |v|^2) = \frac{1}{2}$$

with equality if and only if $|u| = |v| = 2^{-1/2}$. Therefore, the supremum is $\frac{1}{2}$ in this case.

Comment. If V is a real, rather than complex, vector space, then a trigonometric solution is possible. Wolog, we can assume that a basis has been selected so that $z_1 = (1, 0)$ and $z_2 = (\cos\theta, \sin\theta)$. Suppose that $z = (\cos\phi, \sin\phi)$, then

$$\langle z, z_1 \rangle \langle z, z_2 \rangle = \cos\phi(\cos\theta\cos\phi + \sin\theta\sin\phi) = \cos\phi(\cos(\theta - \phi))$$

$$= \frac{1}{2}[\cos\theta + \cos(\theta - 2\phi)].$$

If $\cos\theta = 0$ and $\sin\theta = \pm 1$, then the value of this expression is $|\frac{1}{2}\cos(\theta - 2\phi)| \leq \frac{1}{2}$, so that the supremum is equal to $1/2$, attainable when $2\phi = \theta$.

On the other hand, when $\cos \theta > 0$ and $2\phi = \theta$, the expression is greater than $1/2$ and the supremum exceeds $1/2$. Similarly, when $\cos \theta < 0$, and $2\phi = \theta + \pi$, the expression is less than $-1/2$ and again the supremum exceeds $1/2$.

2013:8. For any real square matrix A, the adjugate matrix, adj A, has as its elements the cofactors of the transpose of A, so that

$$A \cdot \text{adj } A = \text{adj } A \cdot A = (\det A)I.$$

(a) Suppose that A is an invertible square matrix. Show that

$$(\text{adj } (A^t))^{-1} = (\text{adj } (A^{-1}))^t.$$

(b) Suppose that adj (A^t) is orthogonal (i.e., its inverse is its transpose). Prove that A is invertible.

(c) Let A be an invertible $n \times n$ square matrix and let $\det (tI - A) = t^n + c_1 t^{n-1} + \cdots + c_{n-1} t + c_n$ be the characteristic polynomial of the matrix A. Determine the characteristic polynomial of adj A.

Solution. (a) Since $A^t \text{adj } (A^t) = (\det A)I$ and $A^{-1}(\text{adj } (A^{-1})) = (\det A)^{-1} \cdot I$, it follows that

$$(\text{adj } (A^t))^{-1} = \frac{1}{\det A} A^t = (\text{adj } (A^{-1}))^t.$$

(b) Since adj $(A^t) \cdot (\text{adj } (A^t))^t = I$, then

$$A^t = A^t \text{adj } (A^t)(\text{adj } A^t)^t = (\det A)(\text{adj } A^t)^t.$$

Since $A \neq O$, $\det A \neq 0$.

(c)

$$\det(tI - \text{adj } A) = \frac{1}{\det A} \cdot \det(tA - A \text{ adj } A) = \frac{1}{\det A} \cdot \det(tA - (\det A)I)$$

$$= \frac{(-t)^n}{\det A} \cdot \det \left(\frac{\det A}{t} I - A \right)$$

$$= \frac{(-t)^n}{(-1)^n c_n} \left(\sum_{k=0}^{n} \left(\frac{(-1)^n c_n}{t} \right)^{n-k} c_k \right)$$

$$= \sum_{k=0}^{n} (-1)^{n(n-k)} c_k c_n^{n-k-1} t^k,$$

with $c_0 = 1$.

2014:1. The *permanent*, per A, of an $n \times n$ matrix $A = (a_{i,j})$, is equal to the sum of all possible products of the form $a_{1,\sigma(1)} a_{2,\sigma(2)} \cdots a_{n,\sigma(n)}$, where σ runs over all the permutations on the set $\{1, 2, \ldots, n\}$. (This is similar to the definition of determinant, but there is no sign factor.) Show that, for any $n \times n$ matrix $A = (a_{i,j})$ with positive real terms,

$$\text{per } A \geq n! \left(\prod_{1 \leq i,j \leq n} a_{i,j} \right)^{\frac{1}{n}}.$$

Solution. By the arithmetic-geometric means inequality, the sum defining the permanent (having $n!$ terms) is not less than $n!$ times the product of all these terms raised to the power $1/n!$. Each term in the definition of the permanent has n factors, so that product of all the terms has $n \cdot n!$ factors. There are n^2 entries $a_{i,j}$; each appears the same number $n \times n! \div n^2 = (n-1)!$ times. The result follows.

2015:9. What is the dimension of the vector subspace of $\mathbb{R}^n$ generated by the set of vectors

$$(\sigma(1), \sigma(2), \sigma(3), \ldots, \sigma(n))$$

where σ runs through all $n!$ of the permutations of the first n natural numbers.

Solution 1. [J. Love] The dimension cannot exceed n. Taking the difference of two permutations with identical outcomes except in the ith and nth positions where σ takes the values 1 and 2, we find that the vector space contains the vectors

$$(0, 0, \ldots, 0, 1, 0, \ldots, 0, -1) = (3, 4, \ldots, i+1, 2, i+2, \ldots, n, 1)$$
$$- (3, 4, \ldots, i+1, 1, i+2, \ldots, n, 2).$$

This set S of $n-1$ vectors is linearly independent and generates the $(n-1)$-dimensional subspace $\{(x_1, x_2, \ldots, x_n) : x_1 + x_2 + \cdots + x_n = 0\}$. However, the sum of the entries of any element in the generating set is $\frac{1}{2}n(n+1)$, so that any generators along with S is a basis of n elements for the whole space. Thus the required dimension is n.

Solution 2. [C. Wang] As in the first solution, we can show that the span of the generating set contains all vectors of the form $(1, 0, \ldots, -1, \ldots, 0)$ and therefore the vector

$$\frac{1}{2}n(n+1)(1, 0, 0, \ldots, 0, 0) = (1, 2, 3, \ldots, n-1, n) + (2, -2, 0, \ldots, 0, 0)$$
$$+ (3, 0, -3, \ldots, 0, 0) + (n-1, 0, 0, \ldots, -(n-1), 0)$$
$$+ (n, 0, 0, \ldots, 0, -n).$$

Similarly, it can be shown that this span contains all of the basis vectors $(0, 0, \ldots, 1, \ldots, 0)$. Hence the required dimension is n.

Solution 3. The dimension cannot exceed n, and is in fact equal to n. We prove that

$$\{(1, 2, 3, 4, \cdots, n-1, n), (2, 3, 4, 5, \cdots, n, 1), (3, 4, 5, 6, \cdots, 1, 2), \cdots,$$
$$(n, 1, 2, 3, \cdots, n-2, n-1)\}$$

is linearly independent by showing that the determinant of the matrix whose rows are these vectors is nonzero.

Using the fact that the absolute value of the determinant remains unchanged if one column or row is subtracted from another, we have that

$$
\begin{vmatrix}
1 & 2 & 3 & 4 & \cdots & n-2 & n-1 & n \\
2 & 3 & 4 & 5 & \cdots & n-1 & n & 1 \\
3 & 4 & 5 & 6 & \cdots & n & 1 & 2 \\
 & \cdots & & \cdots & & & \cdots & \\
n & 1 & 2 & 3 & \cdots & n-3 & n-2 & n-1
\end{vmatrix}
$$

$$
=
\begin{vmatrix}
1 & 1 & 1 & 1 & \cdots & 1 & 1 & 1 \\
2 & 1 & 1 & 1 & \cdots & 1 & 1 & -(n-1) \\
3 & 1 & 1 & 1 & \cdots & 1 & -(n-1) & 1 \\
4 & 1 & 1 & 1 & \cdots & -(n-1) & 1 & 1 \\
 & & & \cdots & & & \cdots & \\
n & -(n-1) & 1 & 1 & \cdots & 1 & 1 & 1
\end{vmatrix}
$$

$$
=
\begin{vmatrix}
1 & 1 & 1 & 1 & \cdots & 1 & 1 & 1 \\
1 & 0 & 0 & 0 & \cdots & 0 & 0 & -n \\
2 & 0 & 0 & 0 & \cdots & 0 & -n & 0 \\
3 & 0 & 0 & 0 & \cdots & -n & 0 & 0 \\
 & & & \cdots & & & \cdots & \\
n-1 & -n & 0 & 0 & \cdots & 0 & 0 & 0
\end{vmatrix}
$$

$$
= n^{n-2}
\begin{vmatrix}
n & 1 & 1 & 1 & \cdots & 1 & 1 & 1 \\
1 & 0 & 0 & 0 & \cdots & 0 & 0 & -1 \\
2 & 0 & 0 & 0 & \cdots & 0 & -1 & 0 \\
3 & 0 & 0 & 0 & \cdots & -1 & 0 & 0 \\
 & & & \cdots & & & \cdots & \\
n-1 & -1 & 0 & 0 & \cdots & 0 & 0 & 0
\end{vmatrix}
$$

$$
= n^{n-1}
\begin{vmatrix}
n(n+1)/2 & 1 & 1 & 1 & \cdots & 1 & 1 & 1 \\
0 & 0 & 0 & 0 & \cdots & 0 & 0 & -1 \\
0 & 0 & 0 & 0 & \cdots & 0 & -1 & 0 \\
0 & 0 & 0 & 0 & \cdots & -1 & 0 & 0 \\
 & & & \cdots & & & \cdots & \\
0 & -1 & 0 & 0 & \cdots & 0 & 0 & 0
\end{vmatrix}
$$

$$
= n^{n-2} \cdot \frac{n(n+1)}{2} \cdot (1) \neq 0.
$$

CHAPTER 8

Geometry

2001:2. Let $O = (0,0)$ and $Q = (1,0)$. Find the point P on the line with equation $y = x + 1$ for which the angle OPQ is a maximum.

The solution to this problem appears in Chap. 5.

2001:8. A regular heptagon (polygon with seven equal sides and seven equal angles) has diagonals of two different lengths. Let a be the length of a side, b be the length of a shorter diagonal and c be the length of a longer diagonal of a regular heptagon (so that $a < b < c$). Prove ONE of the following relationships:

$$\frac{a^2}{b^2} + \frac{b^2}{c^2} + \frac{c^2}{a^2} = 6$$

or

$$\frac{b^2}{a^2} + \frac{c^2}{b^2} + \frac{a^2}{c^2} = 5.$$

The solution to this problem appears in Chap. 2.

2002:1. Let A, B, C be three pairwise orthogonal faces of a tetrahedron meeting at one of its vertices and having respective areas a, b, c. Let the face D opposite this vertex have area d. Prove that

$$d^2 = a^2 + b^2 + c^2 \ .$$

Solution 1. Let the tetrahedron be bounded by the three coordinate planes in $\mathbb{R}^3$ and the plane with equation $\frac{x}{u} + \frac{y}{v} + \frac{z}{w} = 1$, where u, v, w are positive. The vertices of the tetrahedron are $(0,0,0)$, $(u,0,0)$, $(0,v,0)$, $(0,0,w)$. Let d, a, b, c be the areas of the faces opposite these respective vertices. Then the volume V of the tetrahedron is equal to

$$\frac{1}{3}au = \frac{1}{3}bv = \frac{1}{3}cw = \frac{1}{3}dk,$$

© Springer International Publishing Switzerland 2016
E.J. Barbeau, *University of Toronto Mathematics Competition*
(2001–2015), Problem Books in Mathematics,
DOI 10.1007/978-3-319-28106-3_8

where k is the distance from the origin to its opposite face. The foot of the perpendicular from the origin to this face is located at $((um)^{-1}, (vm)^{-1}, (wm)^{-1})$, where $m = u^{-2} + v^{-2} + w^{-2}$, and its distance from the origin is $m^{-1/2}$. Since $a = 3Vu^{-1}$, $b = 3Vv^{-1}$, $c = 3Vw^{-1}$ and $d = 3Vm^{1/2}$, the result follows.

Solution 2. [J. Chui] Let edges of lengths x, y, z be common to the respective pairs of faces of areas (b, c), (c, a), (a, b). Then $2a = yz$, $2b = zx$ and $2c = xy$. The fourth face is bounded by sides of length $u = \sqrt{y^2 + z^2}$, $v = \sqrt{z^2 + x^2}$ and $w = \sqrt{x^2 + y^2}$. By Heron's formula, its area d is given by the relation

$$
\begin{aligned}
16d^2 &= (u + v + w)(u + v - w)(u - v + w)(-u + v + w) \\
&= [(u + v)^2 - w^2][w^2 - (u - v)^2] \\
&= [2uv + (u^2 + v^2 - w^2)][2uv - (u^2 + v^2 - w^2)] \\
&= 2u^2v^2 + 2v^2w^2 + 2w^2u^2 - u^4 - v^4 - w^4 \\
&= 2(y^2 + z^2)(x^2 + z^2) + 2(x^2 + z^2)(x^2 + y^2) + 2(x^2 + y^2)(y^2 + z^2) \\
&\quad - (y^2 + z^2)^2 - (x^2 + z^2)^2 - (x^2 + y^2)^2 \\
&= 4x^2y^2 + 4x^2z^2 + 4y^2z^2 = 16a^2 + 16b^2 + 16c^2,
\end{aligned}
$$

whence the result follows.

Solution 3. Use the notation of Solution 2. There is a plane containing the edge bounding the faces of areas a and b perpendicular to the edge bounding the faces of areas c and d. Suppose it cuts the latter faces in altitudes of respective lengths u and v. Then $2c = u\sqrt{x^2 + y^2}$, whence $u^2(x^2 + y^2) = x^2y^2$. Hence

$$
v^2 = z^2 + u^2 = \frac{x^2y^2 + x^2z^2 + y^2z^2}{x^2 + y^2} = \frac{4(a^2 + b^2 + c^2)}{x^2 + y^2},
$$

so that

$$
2d = v\sqrt{x^2 + y^2} \implies 4d^2 = 4(a^2 + b^2 + c^2),
$$

as desired.

Solution 4. [R. Ziman] Let $\mathbf{a}$, $\mathbf{b}$, $\mathbf{c}$, $\mathbf{d}$ be vectors orthogonal to the respective faces of areas a, b, c, d that point inwards from these faces and have respective magnitudes a, b, c, d. If the vertices opposite the respective faces are $\mathbf{x}$, $\mathbf{y}$, $\mathbf{z}$, $\mathbf{O}$, then the first three are pairwise orthogonal and

$$
2\mathbf{a} = \mathbf{y} \times \mathbf{z}, \qquad\qquad 2\mathbf{b} = \mathbf{z} \times \mathbf{x}, \qquad\qquad 2\mathbf{c} = \mathbf{x} \times \mathbf{y},
$$

and

$$
2\mathbf{d} = (\mathbf{z} - \mathbf{y}) \times (\mathbf{z} - \mathbf{x}) = -(\mathbf{z} \times \mathbf{x}) - (\mathbf{y} \times \mathbf{z}) - (\mathbf{x} \times \mathbf{y}).
$$

Hence $\mathbf{d} = -(\mathbf{a} + \mathbf{b} + \mathbf{c})$, so that

$$
d^2 = \mathbf{d} \cdot \mathbf{d} = (\mathbf{a} + \mathbf{b} + \mathbf{c}) \cdot (\mathbf{a} + \mathbf{b} + \mathbf{c}) = a^2 + b^2 + c^2.
$$

2004:9. Let $ABCD$ be a convex quadrilateral for which all sides and diagonals have rational length and AC and BD intersect at P. Prove that AP, BP, CP, DP all have rational length.

Solution 1. Because of the symmetry, it is enough to show that the length of AP is rational. The rationality of the lengths of the remaining segments can be shown similarly. Coordinatize the situation by taking $A \sim (0,0)$, $B \sim (p,q)$, $C \sim (c,0)$, $D \sim (r,s)$ and $P \sim (u,0)$. Then, equating slopes, we find that

$$\frac{s}{r-u} = \frac{s-q}{r-p}$$

so that

$$\frac{sr-ps}{s-q} = r-u$$

whence $u = r - \frac{sr-ps}{s-q} = \frac{ps-qr}{s-q}$.

Note that $|AB|^2 = p^2 + q^2$, $|AC|^2 = c^2$, $|BC|^2 = (p^2 - 2pc + c^2) + q^2$, $|CD|^2 = (c^2 - 2cr + r^2) + s^2$ and $|AD|^2 = r^2 + s^2$, we have that

$$2rc = |AC|^2 + |AD|^2 - |CD|^2$$

so that, since c is rational, r is rational. Hence s^2 is rational.

Similarly

$$2pc = |AC|^2 + |AB|^2 - |BC|^2.$$

Thus, p is rational, so that q^2 is rational.

$$2qs = q^2 + s^2 - (q-s)^2 = q^2 + s^2 - [(p-r)^2 + (q-s)^2] + p^2 - 2pr + r^2$$

is rational, so that both qs and $q/s = (qs)/s^2$ are rational. Hence

$$u = \frac{p - r(q/s)}{1 - (q/s)}$$

is rational.

Solution 2. By the cosine law, the cosines of all of the angles of the triangles ACD, BCD, ABC and ABD are rational. Now

$$\frac{AP}{AB} = \frac{\sin \angle ABP}{\sin \angle APB}$$

and

$$\frac{CP}{BC} = \frac{\sin \angle PBC}{\sin \angle CPB}.$$

Since $\angle APB + \angle BPC = 180°$, therefore $\sin \angle APB = \sin \angle CPB$ and

$$\begin{aligned}
\frac{AP}{CP} &= \frac{AB \sin \angle ABP}{BC \sin \angle PBC} = \frac{AB \sin \angle ABP \sin \angle PBC}{BC \sin^2 \angle PBC} \\
&= \frac{AB(\cos \angle ABP \cos \angle PBC - \cos(\angle ABP + \angle PBC))}{BC(1 - \cos^2 \angle PBC)} \\
&= \frac{AB(\cos \angle ABD \cos \angle DBC - \cos \angle ABC)}{BC(1 - \cos^2 \angle DBC)}
\end{aligned}$$

is rational. Also $AP + CP$ is rational, so that $(AP/CP)(AP + CP) = ((AP/CP) + 1)AP$ is rational. Hence AP is rational.

2006:4. Two parabolas have parallel axes and intersect in two points. Prove that their common chord bisects the segments whose endpoints are the points of contact of their common tangent.

Solution 1. Wolog, we may assume that the parabolas have the equations $y = ax^2$ and $y = b(x-1)^2 + c$. The common chord has equation

$$a[b(x-1)^2 + c - y] - b[ax^2 - y] = 0,$$

or

(8.1) $(a-b)y + 2abx - a(b+c) = 0.$

Consider a point (u, au^2) on the first parabola. The tangent at this point has equation $y = 2aux - au^2$. The abscissa of the intersection point of this tangent with the parabola of equation $y = b(x-1)^2 + c$ is given by the equation

$$bx^2 - 2(b+au)x + (au^2 + b + c) = 0.$$

This has coincident roots if and only if

(8.2) $(b+au)^2 = b(au^2 + b + c) \iff a(a-b)u^2 + 2abu - bc = 0.$

In this situation, the coincident roots are $x = 1 + (au)/b$ and the point of contact of the common tangent with the second parabola is

$$\left(1 + \frac{au}{b}, \frac{a^2u^2}{b} + c\right).$$

The midpoint of the segment joining the two contact points is

$$\left(\frac{b + au + bu}{2b}, \frac{abu^2 + a^2u^2 + bc}{2b}\right).$$

Plugging this into the left side of (8.1) and using (8.2) yields that

$$[1/(2b)][(a-b)a(a+b)u^2 + (a-b)bc + 2ab^2 + 2ab(a+b)u - 2ab(b+c)]$$
$$= [(a+b)/(2b)][a(a-b)u^2 + 2abu - bc] = 0.$$

Thus, the coordinates of the midpoint of the segment satisfy (8.1) and the result follows.

Solution 2. [A. Feizmohammadi] Let the two parabolas have equations $y = ax(x-u)$ and $y = bx(x-v)$ with $au \neq bv$. Since the two parabolas must open the same way for the situation to occur, wolog, we may suppose that $a > b > 0$. The parabolas intersect at the points $(0,0)$ and $((au - bv)/(a-b), (ab(au-bv)(u-v))/(a-b)^2)$, and the common chord has equation $(a-b)y - ab(u-v)x = 0$.

Let $y = mx + k$ be the equation of the common tangent. Then both of the equations $ax^2 - (au+m)x - k = 0$ and $bx^2 - (bv+m)x - k = 0$ have coincident solutions. Therefore $(au+m)^2 + 4ak = (bv+m)^2 + 4bk = 0$, from which (by eliminating k),

$$ab(au^2 - bv^2) + 2ab(u-v)m + (b-a)m^2 = 0.$$

The common tangent of equation $y = mx + k$ touches the first parabola at

$$\left(\frac{au + m}{2a}, \frac{m^2 - a^2 u^2}{4a} \right)$$

and the second parabola at

$$\left(\frac{bv + m}{2b}, \frac{m^2 - b^2 v^2}{4b} \right).$$

The midpoint of the segment joining these two points is

$$\left(\frac{ab(u + v) + (a + b)m}{4ab}, \frac{(a + b)m^2 - ab(au^2 + bv^2)}{8ab} \right).$$

Using these coordinates as the values of x and y, we find that

$$8ab[(a - b)y - ab(u - v)x] = (a - b)[(a + b)m^2 - ab(au^2 + bv^2)]$$
$$- [2a^2b^2(u - v)(u + v) + 2ab(a + b)(u - v)m]$$
$$= (a + b)[(a - b)m^2 - 2ab(u - v)m]$$
$$- [ab(a - b)(au^2 + bv^2) + 2a^2b^2(u^2 - v^2)]$$
$$= (a + b)ab(au^2 - bv^2)$$
$$- ab(a + b)(au^2 - bv^2) = 0.$$

Comment. There are two possible common tangents, and the common chord bisects the segment on each of them (Fig. 8.1).

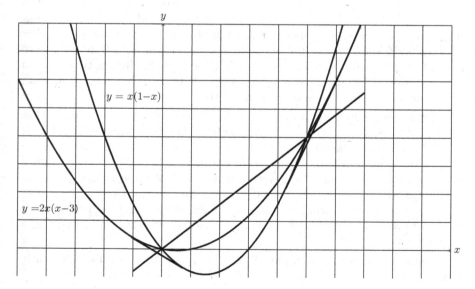

Fig. 8.1. Two parabolas $y = 2x(x - 3)$ and $y = x(x - 1)$ and their common tangents

2006:10. Let P be a planar polygon that is not convex. The vertices can be classified as either *convex* or *concave* according as to whether the angle at the vertex is less than or greater than $180°$ respectively. There must be at least two convex vertices. Select two consecutive convex vertices (i.e., two interior angles less than $180°$ for which all interior angles in between exceed $180°$) and join them by a segment. Reflect the edges between these two convex angles in the segment to form along with the other edges of P a polygon P_1. If P_1 is not convex, repeat the process, reflecting some of the edges of P_1 in a segment joining two consecutive convex vertices, to form a polygon P_2. Repeat the process. Prove that, after a finite number of steps, we arrive at a polygon P_n that is convex.

Solution. Note that for the problem to work, we need to assume that, at each stage, we create a polygon that does not cross itself. Suppose that P is a m-gon. Then each P_i has at most m sides, as some angles of $180°$ may be created at the vertices. The perimeter of P_i is the same as that of P for each index i. Also $P \subseteq P_1 \subseteq P_2 \subseteq \cdots \subseteq P_i \subseteq \cdots$. Therefore there is a disc that contains all of the polygons P_i (the perimeter is at least as long as any line segment contained within the polygon).

Let a be a vertex of P, and let a_i be the corresponding vertex obtained from a in P_i, each a_i being equal to its predecessor or obtained from it by a reflection that takes it outside of P_{i-1}. In the latter case, a_i must lie outside of P_j when $j < i$. Because the sequence $\{a_i\}$ is bounded, it must have an accumulation point a'.

Let $\epsilon > 0$ be small enough that an ϵ-neighbourhood about a excludes all other vertices of P. Consider an ϵ-neighbourhood U of a'. It contains a point a_l. If $a_{l+1} \notin U$, then a_{l+1} is the reflection of a_l in a line joining two other vertices of P_l and lying on the opposite side of the centre of U to a_l, and so $U \subseteq P_{l+1}$. This entails that $a_k \notin U$ for $k \geq l+1$, contradicting that a' is an accumulation point. Hence $a_k \in U$ for $k \geq l$. It follows that $\lim_{i \to \infty} a_i = a'$. We show, in fact, that $a_i = a'$ for sufficiently large i.

Let P' be the limiting polygon with vertices a'. Suppose, if possible that a' is a concave vertex. We can find a neighbourhood $V_{x'}$ around each vertex x' of P' such that, selecting any point from each of these neighbourhoods $V_{x'}$ will create a polygon with a concave vertex in $V_{a'}$. Thus, there is an index r such that the polygon P_r has a vertex inside each of these neighbourhoods. If the concave vertex a_r is reflected about a line joining two other vertices in P_r to a_{r+1}, then a_r will be an interior point of P_{r+1} and each subsequent a_i will lie outside of the interior of P_{r+1}, and so outside of $U_{a'}$, yielding a contradiction.

Hence P' has acute angles at all of its vertices. But then for sufficiently large n, so will P_n and so $a_i = a_n$ for $i \geq n$. The result follows.

2007:8. Suppose that n points are given in the plane, not all collinear. Prove that there are at least n distinct straight lines that can be drawn through pairs of the points.

Solution. This can be obtained as a corollary to Sylvester's Theorem: *Suppose that n points are given in the plane, not all collinear. Then there is exists a line that contains exactly two of them.* The proof of this is given in Appendix A.

We can now solve the problem by an induction argument. It is clearly true for $n = 3$; suppose it holds for $n - 1 \geq 3$ and that n points are given. Pick a line that passes through exactly two points P and Q of the set. At least one of these points, say P, is not collinear with the rest. Remove this line and the point Q. We can find $n - 1$ distinct lines determined by pairs of the other $n - 1$ points, and restoring the line through PQ yields the nth line.

2008:1. Three angles of a heptagon (7-sided polygon) inscribed in a circle are equal to $120°$. Prove that at least two of its sides are equal.

Solution. Consider two adjacent sides of the heptagon for which the angle between them is $120°$. The chord of the circle joining the endpoints that are not common to the chords subtends an angle of $60°$ at the circumference of the circle and therefore an angle of $120°$ at the centre of the circle. If the three pairs of adjacent sides forming angles of $120°$ are mutually disjoint from each other, then they constitute six sides of the heptagon which subtend in total angles totalling $360°$ at the centre of the circle, leaving no positive angle for the seventh side to subtend. Hence, two of the pairs of sides must have an edge in common. However, since each pair subtends the same angle at the centre, the edges that the pairs do not have in common must be of equal length and the result follows.

2008:6. 2008 circular coins, possibly of different diameters, are placed on the surface of a flat table in such a way that no coin is on top of another coin. What is the largest number of points at which two of the coins could be touching?

The solution to this problem appears in Chap. 10.

2008:10. A point is chosen at random (with the uniform distribution) on each side of a unit square. What is the probability that the four points are the vertices of a quadrilateral with area exceeding $\frac{1}{2}$?

Solution 1. The desired probability is $\frac{1}{2}$.

Suppose that the points are located in a counterclockwise direction distance x, y, z, w from the four vertices in order. Then the area of that part of the square lying outside of the inner quadrilateral is

$$\frac{1}{2}[x(1-w)+y(1-x)+z(1-y)+w(1-z)] = \frac{1}{2}[(x+y+z+w)-(xw+yx+zy+wz)].$$

The condition that the area of the inner quadrilateral exceeds $\frac{1}{2}$ is that

$$(x + y + z + w) - (xw + yx + zy + wz) < 1.$$

The area is equal to $\frac{1}{2}$ if and only if $x + z = 1$ or $y + w = 1$, both of which occur with probability 0.

On the other hand, note that the points are located at distances $x, w' = 1 - w$ from one vertex of the square and distance $z, y' = 1 - y$ from the diagonally opposite vertex. Then the area of that part of the square lying outside the inner quadrilateral is

$$\frac{1}{2}[xw' + (1-x)(1-y') + zy' + (1-w')(1-z)] = \frac{1}{2}[(xw' + xy' + zy' + zw') - (x + y' + z + w') + 2],$$

so that the condition that the area of the inner quadrilateral exceeds $\frac{1}{2}$ is that

$$(x + y' + z + w') - (xw' + y'x + zy' + w'z) > 1.$$

Since x, y, z, w and so x, y', z, w', are chosen independently and uniformly in each case, the probabilities of either inequality occurring is the same as the probability of the other. Therefore, the desired probability is $\frac{1}{2}$.

Solution 2. Let the square have vertices $(0,0)$, $(1,0)$, $(1,1)$, $(0,1)$ in the cartesian plane and let the four selected points by $(a,0)$, $(1,b)$, $(c,1)$ and $(0,d)$, where $0 \leq a, b, c, d \leq 1$. Then the area of the inner quadrilateral is equal to

$$1 - \frac{1}{2}(1-a)b - \frac{1}{2}(1-c)(1-b) - \frac{1}{2}(1-d)c - \frac{1}{2}ad = \frac{1}{2}[1 + (d-b)(c-a)].$$

The area of the quadrilateral is exactly $\frac{1}{2}$ when $b = d$ or $a = c$, which occurs with probability 0. The area exceeds $\frac{1}{2}$ if and only if, either $b < d$ and $a < c$, or $b > d$ and $a > c$. The area is less than $\frac{1}{2}$ if and only if, either $b < d$ and $c < a$, or $b > d$ and $c > a$. Since a, b, c, d are selected randomly and independently from $[0, 1]$, by symmetry, the two latter events have equal probability of $\frac{1}{2}$.

2010:1. Let F_1 and F_2 be the foci of an ellipse and P be a point in the plane of the ellipse. Suppose that G_1 and G_2 are points on the ellipse for which PG_1 and PG_2 are tangents to the ellipse. Prove that $\angle F_1 P G_1 = \angle F_2 P G_2$.

Solution. Let H_1 be the reflection of F_1 in the tangent PG_1, and H_2 be the reflection of F_2 in the tangent PG_2. We have that $PH_1 = PF_1$ and $PF_2 = PH_2$. By the reflection property, $\angle PG_1 F_2 = \angle F_1 G_1 Q = \angle H_1 G_1 Q$, where Q is a point on PG_1 produced. Therefore, $H_1 F_2$ intersects the ellipse in G_1. Similarly, $H_2 F_1$ intersects the ellipse in G_2. Therefore

$$H_1 F_2 = H_1 G_1 + G_1 F_2 = F_1 G_1 + G_1 F_2$$
$$= F_1 G_2 + G_2 F_2 = F_1 G_2 + G_2 H_2 = H_2 F_1.$$

Therefore, triangle PH_1F_2 and PF_1H_2 are congruent (SSS), so that $\angle H_1PF_2 = \angle H_2PF_1$. It follows that

$$2\angle F_1PG_1 = \angle H_1PF_1 = \angle H_1PF_2 \pm \angle F_2PF_1$$
$$= \angle H_2PF_1 \pm \angle F_2PF_1 = \angle H_2PF_2 = 2\angle F_2PG_2$$

and the desired result follows.

2012:1. An equilateral triangle of side length 1 can be covered by five equilateral triangles of side length u. Prove that it can be covered by four equilateral triangles of side length u. (A triangle is a closed convex set that contains its three sides along with its interior.)

Solution. Consider the set of six points consisting of the vertices of the triangle and the midpoints of the three sides. By the Pigeonhole Principle, one of the five covering equilateral triangles must cover two of these six points. Since the distance between any two of the six points is at least $\frac{1}{2}$ and any two points within an equilateral triangle cannot be further apart than two of its vertices, we must have that $u \geq \frac{1}{2}$.

However, it is clear that the given equilateral triangle can be covered by four such triangles of side-length $\frac{1}{2}$ determined by joining pairs of midpoints of the sides. Each of these four triangles can be included within a triangle of side-length u.

2012:5. Let $\mathfrak{C}$ be a circle and Q a point in the plane. Determine the locus of the centres of those circles that are tangent to $\mathfrak{C}$ and whose circumference passes through Q.

Solution 1. There are several cases to consider. Let O be the centre and r the radius of the given circle. Let P be the centre of the variable circle.

(i) Q lies on $\mathfrak{C}$. In this case, the variable circle is tangent to $\mathfrak{C}$ at Q and the locus is the set of points on the line OQ with the exception of Q itself.

(ii) Q lies in the interior of $\mathfrak{C}$. Suppose that the variable circle is tangent to $\mathfrak{C}$ at T. Then

$$|OP| + |PO| = |OP| + |PT| = |OT| = r$$

so that P lies on the ellipse with foci O and Q and major axis of length r. (If $Q = O$, this becomes a circle with diameter r.)

Conversely, if P is a point on the ellipse just described, let OP produced meet $\mathfrak{C}$ at T and construct a circle with centre P and radius $|PT|$. This circle will be tangent to $\mathfrak{C}$ and $|QP| = r - |OP| = |TP|$, so that it will pass through Q.

(iii) Q lies outside of $\mathfrak{C}$ and the variable circle touches $\mathfrak{C}$ at T, say, with the two circles external to each other. Let a be the radius of the variable circle. The segment OP contains T and

$$|OP| = r + a = r + |PQ|$$

so that $|OP| - |PQ| = r$, a constant. Hence P lies on one branch of a hyperbola with foci O, Q whose asymptotes are the lines that pass through O and the contact points of the tangents drawn to $\mathfrak{C}$ from Q.

(iv) Q lies outside of $\mathfrak{C}$ and the variable circle touches $\mathfrak{C}$ at T, say, with $\mathfrak{C}$ contained inside the variable circle. Then, with a the radius of the variable circle,

$$|QP| - |OP| = a - (a - r) = r$$

so that P lines on a branch of the hyperbola with foci O, Q.

Solution 2. Let $\mathfrak{C}$ have equation $x^2 + y^2 = 1$ and $Q \sim (q, 0)$, with $q \geq 0$. Suppose that (X, Y) is the centre of a circle that passes through Q and touches $\mathfrak{C}$. Then

$$\sqrt{(X-q)^2 + Y^2} = |\sqrt{X^2 + Y^2} - 1|$$
$$\implies X^2 - 2qX + q^2 + Y^2 = X^2 + Y^2 + 1 - 2\sqrt{X^2 + Y^2}$$
$$\implies 2\sqrt{X^2 + Y^2} = (1 - q^2) + 2qX$$
$$\implies 4X^2 + 4Y^2 = (1 - q^2)^2 + 4q(1 - q^2)X + 4q^2X^2$$
$$\implies (1 - q^2)(2X - q)^2 + 4Y^2 = (1 - q^2).$$

When $q = 0$, the locus is a circle concentric with the given circle with half the radius. When $0 < q < 1$, the locus is an ellipse with centre at $(q/2, 0)$, semi-major axis $1/2$ and semi-minor axis $\sqrt{1 - q^2}/2$. When $q = 1$, the locus is the line $Y = 0$.

Suppose that $q > 1$. The equation of the locus can be written

$$(2X - q)^2 - \frac{4Y^2}{q^2 - 1} = 1,$$

from which it can be seen that the locus is a hyperbola with asymptotes given by the equations

$$\sqrt{q^2 - 1}(2X - q) = \pm 2Y.$$

These are lines passing through the point $(q/2, 0)$ with slopes $\pm\sqrt{q^2 - 1}$.

The tangents to the circle $\mathfrak{C}$ from $(q, 0)$ touch the circle at the points $(1/q, \pm\sqrt{q^2 - 1}/q)$. When the centre of the variable circle is on the right branch of the hyperbola, the two circles touch externally at a point that lies on the minor arc between these points. When the centre is on the left branch of the hyperbola, $\mathfrak{C}$ touches the variable circle internally at a point on the major arc joining these two points. Note that the vertex of the left branch of the hyperbola is $((q - 1)/2, 0)$ which lies on the positive x-axis; this is the centre of a circle of radius $(q + 1)/2 > 1$ that touches $\mathfrak{C}$ at the point $(-1, 0)$.

2012:8. Determine the area of the set of points (x, y) in the plane that satisfy the two inequalities:

$$x^2 + y^2 \le 2$$
$$x^4 + x^3 y^3 \le xy + y^4.$$

Solution. The second inequality can be rewritten as

$$0 \le y^4 + xy - x^4 - x^3 y^3 = (y^3 + x)(y - x^3).$$

The locus of the equation $y^3 + x = 0$ can be obtained from the locus of the equation $y - x^3 = 0$ by a $90°$ rotation about the origin that relates the points (u, v) and $(v, -u)$. Thus the two curves partition the disc $x^2 + y^2 \le 2$ into four regions of equal area, and the inequality is satisfied by the two regions that cover the y-axis. Thus the area of the set is half the area of the disc, namely π.

2013:2. ABCD is a square; points U and V are situated on the respective sides BC and CD. Prove that the perimeter of triangle CUV is equal to twice the sidelength of the square if and only if $\angle UAV = 45°$.

Solution 1. [Y. Babich; J. Song] Produce CD to W so that $DW = BU$. Then triangles ABU and ADW are congruent (SAS) so that $AU = AW$.

First, assume that $\angle UAV = 45°$. Then

$$\angle VAW = \angle VAD + \angle DAW = \angle VAD + \angle BAU$$
$$= 90° - \angle UAV = 45°.$$

Since also $AU = AW$ and AV is common, triangles UAV and WAV are congruent (SAS) and $UV = VW = DV + BU$.

Therefore

$$CU + CV + UV = (CU + BU) + (CV + DV) = BC + CD$$

as desired.

On the other hand, assume that $CU + CV + UV = BC + CD$. Then

$$UV = (BC - CU) + (CD - CV) = BU + VD = VW,$$

so that triangles AUV and AWV are congruent (SSS) and $\angle UAV = \angle WAV = \angle VAD + \angle UAB$. Since $\angle UAB + \angle UAV + \angle VAD = 90°$, it follows that $\angle UAV = 45°$.

Solution 2. Let the side length of the square be 1, $|BU| = u$, $|DV| = v$, $|UV| = w$. Then $|CU| = 1 - u$ and $|CV| = 1 - v$.

Suppose that $\angle UAV = 45°$. Since $45° = \angle BAU + \angle DAV$,

$$1 = \tan 45° = \frac{u + v}{1 - uv},$$

whereupon $1 - uv = u + v$ and $1 + u^2v^2 = u^2 + 4uv + v^2$. By the Law of Cosines,

$$w^2 = (1 + u^2) + (1 + v^2) - \sqrt{2(1 + u^2)(1 + v^2)}$$

$$= u^2 + v^2 + 2 - \sqrt{2(1 + u^2v^2 + u^2 + v^2)} = u^2 + v^2 + (2 - \sqrt{4(u^2 + 2uv + v^2)})$$

$$= u^2 + v^2 + 2(1 - (u + v)) = u^2 + v^2 + 2uv = (u + v)^2.$$

Hence $w = u + v$ and the perimeter of CUV is equal to $(1 - u) + (1 - v) + (u + v) = 2$.

On the other hand, suppose that $|UV| = u + v$. Then, by Pythagoras' Theorem,

$$u^2 + 2uv + v^2 = (1 - 2u + u^2) + (1 - 2v + v^2)$$

so that $u + v = 1 - uv$. Therefore

$$\tan(\angle UAB + \angle VAD) = \frac{u + v}{1 - uv} = 1,$$

so that $\angle UAB + \angle VAD = 45°$ and $\angle UAV = 45°$.

Solution 3. [Z. Liu] Let $A \sim (0, 0)$, $B \sim (1, 0)$, $C \sim (1, 1)$, $D \sim (0, 1)$, $U \sim (1, u)$, and $V \sim (v, 1)$, where $0 < u, v < 1$. Since

$$\cos \angle UAV \cdot |AU| \cdot |AV| = \overrightarrow{AU} \cdot \overrightarrow{AV},$$

$$\angle UAV = 45° \Leftrightarrow \sqrt{2}(u + v) = \sqrt{1 + u^2}\sqrt{1 + v^2}$$

$$\Leftrightarrow (u + v)^2 = u^2 + 2uv + v^2 = 1 - 2uv + u^2v^2 = (1 - uv)^2$$

$$\Leftrightarrow u + v = 1 - uv \Leftrightarrow (1 - u)^2 + (1 - v)^2 = (u + v)^2$$

$$\Leftrightarrow |UV| = |BU| + |VD|$$

$$\Leftrightarrow |CU| + |CV| + |UV| = |BU| + |CU| + |CV| + |VD| = |BC| + |CD|.$$

The result follows.

Solution 4. [J. Zung] Let U be fixed on BC. Then as V moves from C to D, the lengths of CV and UV strictly increase, so that the perimeter of triangle CUV strictly increases from $2CU < BC + CD$ to $CU + UD + CD > BC + CD$. Therefore, there is a unique position of V such that $CU + CV + UV = BC + CD$.

Also, as V moves from C to D, the angle UAV strictly increases from an angle less than $45°$ to an angle greater than $45°$. Therefore, there is a unique position of V such that $\angle UAV = 45°$. We need to show that the position of V is the same in both situations.

Select V so that UV is tangent to the circle with centre A that passes through B and D. Let T be the point of tangency. Then

$$CU + CV + VU = CU + CV + VT + TU = CU + CV + VD + UB = CD + CB.$$

Also, $\angle VAT = \frac{1}{2}\angle DAT$ and $\angle UAT = \frac{1}{2}\angle BAT$, so that $\angle UAV = 45°$. Thus the point V is the unique point for both of the foregoing situations, and the desired result follows.

Solution 5 (of "if" part). [S. Rumsey] Suppose $\angle UAV = 45°$. Let the image of B reflected in AU be E and the image of D reflected in AV be F. Then

$$\angle EAU + \angle FAV = \angle BAU + \angle DAV = 45° = \angle UAV,$$

so that E and F must fall on the same line through A. Since $AE = AB = AD = AF$, then $E = F$.

Also,

$$\angle AEU = \angle ABU = 90° = \angle ADV = \angle AEV,$$

so that U, E, V are collinear and E lies on UV. Therefore

$$CU + CV + UV = CU + CV + UE + EV = CU + CV + BU + DV = BC + CD,$$

as desired.

2013:5. A point on an ellipse is joined to the ends of its major axis. Prove that the portion of a directrix intercepted by the two joining lines subtends a right angle at the corresponding focus.

Solution 1. Suppose that the focus is at the origin of the Cartesian plane, the directrix is the line $x = 1$ and the eccentricity of the ellipse is e. Then the equation of the ellipse is $\sqrt{x^2 + y^2} = e(1 - x)$, or

$$y^2 = -(1 - e^2)x^2 - 2e^2x + e^2.$$

The major axis is the x-axis and the ellipse intersects this axis at the points $(e/(1 + e), 0)$ and $(-e/(1 - e), 0)$. Let (u, v) be an arbitrary point on the ellipse. Then the lines determined by the point (u, v) and these endpoints respectively intersect the directrix at

$$\left(1, \frac{v}{(1 + e)u - e}\right) \quad \text{and} \quad \left(1, \frac{v}{(1 - e)u + e}\right).$$

The product of the slopes of the segments joining these to the origin is

$$\frac{v^2}{(1 - e^2)u^2 + [e(1 + e) - e(1 - e)]u - e^2} = \frac{-(1 - e^2)u^2 - 2e^2u + e^2}{(1 - e^2)u^2 + 2e^2u - e^2} = -1.$$

Solution 2. [J. Song; T. Xiao] Let the equation of the ellipse be $x^2/a^2 + y^2/b^2 = 1$ and the point $P \sim (u, v)$. The right directrix of the ellipse has equation $x = a^2/c$ and the right focus F is at $(c, 0)$, where $a^2 = b^2 + c^2$.

The equation of the line through $(-a, 0)$ and P is $(u + a)y = v(x + a)$ and this meets the directrix at the point

$$G \sim \left(\frac{a^2}{c}, \frac{av(a + c)}{c(u + a)}\right).$$

The equation of the line through $(a, 0)$ and P is $(u - a)y = v(x - a)$ and this meets the directrix at the point

$$H \sim \left(\frac{a^2}{c}, \frac{av(a - c)}{c(u - a)}\right).$$

Let K be the point $(a^2/c, 0)$ where the directrix and major axis intersect. Because P lies on the ellipse, we have that $a^2v^2 = b^2(a^2 - u^2)$.

There are two possible ways to proceed. Taking the dot product of the vectors $\overrightarrow{FG}$ and $\overrightarrow{FH}$, we obtain that

$$\left(\frac{a^2}{c} - c, \frac{av(a+c)}{c(u+a)}\right) \cdot \left(\frac{a^2}{c} - c, \frac{av(a-c)}{c(u-a)}\right) = \frac{1}{c^2}\left[(a^2 - c^2)^2 - \frac{a^2v^2(a^2 - c^2)}{a^2 - u^2}\right]$$
$$= \frac{a^2 - c^2}{c^2}[a^2 - c^2 - b^2] = 0,$$

so that $\angle GFH = 90°$.

Alternatively, we can use the Law of Cosines to obtain that

$$2|GF||HF|\cos\angle GFH = |FG|^2 + |FH|^2 - |GH|^2$$
$$= 2|FK|^2 + |GK|^2 + |HK|^2 - (|GK| + |HK|)^2$$
$$= 2[|FK|^2 - |GK||HK|]$$
$$= 2\left[\left(\frac{a^2}{c} - c\right)^2 - \frac{a^2v^2(a^2 - c^2)}{c^2(a^2 - u^2)}\right]$$
$$= \frac{2}{c^2}[(a^2 - c^2)^2 - b^2(a^2 - c^2)] = \frac{2(a^2 - c^2)}{c^2}[a^2 - c^2 - b^2] = 0,$$

from which the result follows. (Note that $|u - a| = a - u$ since $u < a$.)

2014:8. The hyperbola with equation $x^2 - y^2 = 1$ has two branches, as does the hyperbola with equation $y^2 - x^2 = 1$. Choose one point from each of the four branches of the locus of $(x^2 - y^2)^2 = 1$ such that area of the quadrilateral with these four vertices is minimized.

Solution. Let N, W, S, E be arbitrary points on the four branches that are symmetric respectively about the positive y-axis (northern), negative x-axis (western), negative y-axis (southern) and positive x-axis (eastern). Consider the diagonal NS of $NWSE$. If the secant through W that is parallel to NS passes through two distinct points of the western branch, then the area $[NWS]$ of triangle NSW can be made smaller by replacing W by the point of tangency to the western branch by a tangent parallel to NS. Similarly, the area $[NSE]$ can be made smaller when the tangent at E is parallel to NS. We can follow a similar argument for the diagonal WE. Thus, for any quadrilateral $NWSE$, we can find a quadrilateral of no larger area where the tangents to the hyperbolae at N and S are parallel to WE and the tangents to the hyperbolae at E and W are parallel to NS. Thus we may assume that $NWSE$ is a parallelogram.

Observe that the slopes of the diagonals WE as well as the tangents to the northern and southern branches range over the open interval $(-1, 1)$, and that the slopes of the diagonals NS and the tangents to the western and eastern branches are similarly related. Each allowable tangent slope occurs exactly twice, the tangents being related by a $180°$ rotation about the origin.

Thus, in the minimization problem, we need consider only parallelograms centred at the origin with the foregoing tangent and diagonal relationship.

Let $N \sim (a, b)$ be any point on the northern branch, where, wolog, we may assume that $0 \leq a < b$ (and $b^2 - a^2 = 1$). The slope of the tangent at N is a/b, so the point E, located on the line with this slope through the origin must be at (b, a). Similarly, $S \sim (-a, -b)$ and $W \sim (-b, -a)$. Thus, $NWSE$ is in fact a rectangle whose sides are perpendicular to the asymptotes of the hyperbola and whose area is

$$|NE| \cdot |ES| = (\sqrt{2})(b - a)(\sqrt{2})(b + a) = 2(b^2 - a^2) = 2.$$

Therefore, the required minimum area is 2.

[Problem **2014:8** was contributed by Robert McCann.]

2015:2. Given $2n$ distinct points in space, the sum S of the lengths of all the segments joining pairs of them is calculated. Then n of the points are removed along with all the segments having at least one endpoint from among them. Prove that the sum of the lengths of all the remaining segments is less than $\frac{1}{2}S$.

Solution 1. [J. Love] Let $P_1, P_2, \ldots, P_n$ be the points that remain after the points $Q_1, Q_2, \ldots, Q_n$ are removed. By repeated application of the triangle inequality, for each of the $\binom{n}{2}$ pairs (i, j), we have that

$$|P_iP_j| \leq |P_iQ_j| + |Q_jQ_i| + |Q_iP_j|.$$

Note that, if $(i, j) \neq (r, s)$, then $P_iP_j \neq P_rP_s$, $P_iQ_j \neq P_rQ_s$ and $Q_iQ_j \neq Q_rQ_s$. For each retained segment, there is a distinct set of three deleted intervals whose total length exceeds that of the retained segment. Hence, the total length of the deleted segments exceeds that of the retained segments. (The inequality is strict because we have not included among the deleted segments those of the form P_iQ_i.) The result follows.

Solution 2. Define P_i and Q_i as before, For each triple i, j, k with $i \neq j$ and $1 \leq i, j, k \leq n$, write the triangle inequality

$$|P_iP_j| \leq |P_iQ_k| + |P_jQ_k|.$$

There are $n\binom{n}{2}$ inequalities in all; for each of the $\binom{n}{2}$ choices of P_iP_j we have an inequality for each of the n choices of Q_k. Each $|P_iP_j|$ appears in n of the inequalities. There are $2n\binom{n}{2} = n^2(n - 1)$ terms on the right side of the inequalities and each of the n^2 terms of the form $|P_iQ_k|$ appears $n - 1$ times. Let T be the sum of all the lengths $|P_iP_j|$. Then the sum of the lengths of all the intervals involving at least one Q_k is $S - T$, and this includes all the intervals of the form P_iQ_k. Adding all the inequalities yields that $nT \leq (n - 1)(S - T)$, from which we find that

$$(2n - 1)T \leq (n - 1)S.$$

The desired result follows.

Solution 3. [S. Xiu] We begin with $2n$ points $X_1, X_2, \ldots, X_{2n}$ and remove the points one by one beginning with X_{2n}. Let S_k be the sum of the lengths of all the segments joining pairs of points in $\{X_1, X_2, \ldots, X_k\}$ for $n \leq k \leq 2n-1$. When $1 \leq i < j \leq k$, $|X_iX_j| \leq |X_iX_{k+1}| + |X_jX_{k+1}|$. There are $\binom{k}{2}$ inequalities of this type in which each $|X_iX_{k+1}|$ appears $k-1$ times, once for each X_iX_j with $i \neq j$, so that $S_k \leq (k-1)\sum_{i=1}^{k} |X_iX_{k+1}|$. Therefore

$$(k-1)S_{k+1} = (k-1)(S_k + \sum_{i=1}^{k} |X_iX_{k+1}|) \geq kS_k.$$

Thus $S_k \leq [(k-1)/k]S_{k+1}$. Hence

$$S_n \leq \frac{n-1}{n} \cdot \frac{n}{n+1} \cdot \ldots \cdot \frac{2n-2}{2n-1}S_{2n} = \frac{n-1}{2n-1}S < \frac{1}{2}S.$$

CHAPTER 9

Group Theory

2003:10. Let G be a finite group of order n. Show that n is odd if and only if each element of G is a square.

Solution. Suppose that n is odd. Then, by Lagrange's Theorem, each element of G is of odd order, so that, if $a \in G$, then $a^{2k+1} = e$ (the identity) for some nonnegative integer k. Hence $a = (a^{k+1})^2$ is a square. On the other hand, suppose that n is even. Pair off each element of G with its inverse. Some elements get paired off with a distinct element, and others get paired off with themselves. Since n is even and an even number of elements get paired off with a distinct element, there must be an even number of elements that get paired off with themselves. Since the identity gets paired off with itself, there must be some other element v that also is its own inverse, i.e. $v^2 = e$. Consider the mapping $x \to x^2$ defined from G to itself. Since e and v have the same image and since G is finite, this mapping cannot be onto. Hence, there must be an element of G which is not the square of another element. The result follows.

2005:6. Let G be a subgroup of index 2 contained in S_n, the group of all permutations of n elements. Prove that $G = A_n$, the alternating group of all even permutations.

Solution 1. Let $u \in S_n \backslash G$. Since G is of index 2, $uG = Gu$ whence $u^{-1}Gu = G$. Thus, $x^{-1}gx \in G$ for each element g in G and x in S_n. If t is a transposition, than $x^{-1}tx$ runs through all transpositions as x runs through G. Therefore, if G contains a transposition, then G must contain every transposition, and so must be all of S_n, giving a contradiction. Thus, every transposition must belong to Gu. Hence, if t_1 and t_2 are two transpositions, then $t_1 = g_1 u$ and $t_2 = t_2^{-1} = g_2 u$ for g_1 and g_2 in G. Hence, $t_1 t_2 = (g_1 u)(u^{-1} g_2^{-1}) = g_1 g_2^{-1} \in G$. Thus, G contains every even permutation, and so contains A_n. The result follows.

E.J. Barbeau, *University of Toronto Mathematics Competition (2001–2015)*, Problem Books in Mathematics,
DOI 10.1007/978-3-319-28106-3_9

Solution 2. As in Solution 1, we see that G must be normal, and so must be the kernel of a homomorphism ϕ from S_n into the multiplicative group $\{1, -1\}$. Suppose that $x \in S_n$ and x has order m, then $1 = \phi(x^m) = (\phi(x))^m$, whence m is even or $x \in G$. It follows that every element of odd order must belong to G, and so in particular, G must contain all 3-cycles. Since $(ab)(ac) = (abc)$ and $(ab)(cd) = (abc)(adc)$ (multiplying from the left), every product of a pair of transpositions and hence every even permutation is a product of 3-cycles. Hence G contains all even permutations and so must coincide with A_n.

2008:7. Let G be a group of finite order and identity e. Suppose that ϕ is an automorphism of G onto itself with the following properties: (1) $\phi(x) = x$ if and only if $x = e$; (2) $\phi(\phi(x)) = x$ for each element x of G.

(a) Give an example of a group and automorphism for which these conditions are satisfied.
(b) Prove that G is commutative (i.e., $xy = yx$ for each pair x, y of elements in G).

Solution. (a) An example is $\mathbb{Z}_3 \equiv \{0, 1, 2\}$, the integers modulo 3 with mod 3 addition as the group operation, with the automorphism that switches 1 and 2.

(b) Suppose for each element x, we denote $\phi(x)$ by x'. We define the mapping $\alpha : G \to G$ by $\alpha(x) = x'x^{-1}$. We show that α is one-one.

Suppose that $\alpha(x) = \alpha(y)$. Then $x'x^{-1} = y'y^{-1}$, so that

$$x^{-1}y = x'^{-1}y' = (x^{-1})'y' = (x^{-1}y)'.$$

Hence $x^{-1}y = e$, so that $y = x$. Since α is one-one and G is finite, α is onto.

For any element $z \in G$, select x so that $z = x'x^{-1}$. Then $z' = x(x^{-1})'$, so that

$$zz' = x'x^{-1}x(x')^{-1} = e = z'z,$$

whence $z' = z^{-1}$. Therefore

$$\phi(z) = z^{-1}$$

for each $z \in G$.

Let x, y be any two elements of G. Then

$$xy = [(xy)']' = (x'y')' = (x^{-1}y^{-1})^{-1} = yx.$$

The desired result follows.

2011:8. The set of transpositions of the symmetric group S_5 on $\{1, 2, 3, 4, 5\}$ is

$$\{(12), (13), (14), (15), (23), (24), (25), (34), (35), (45)\}$$

where (ab) denotes the permutation that interchanges a and b and leaves every other element fixed. Determine a product of all transpositions, each occuring exactly once, that is equal to the identity permutation ϵ, which leaves every element fixed.

Solution. Multiplication proceeds from left to right. Observe that $(ab)(ac)(bc) = (ac)$ for any a, b and c. We observe that $(13)(23)$ carries 1 to 2 and that $(24)(14)$ carries 2 to 1, so that $(13)(24)(14)(23) = (12)(34)$. Therefore

$$(12)(13)(24)(14)(23)(34) = \epsilon.$$

Therefore

$$(15)(12)(25)(13)(24)(14)(23)(35)(34)(45) = \epsilon.$$

Comment. Other answers were

$$(34)(45)(35)(23)(24)(15)(25)(14)(13)(12) \quad \text{(S. Sagatov)},$$
$$(12)(23)(34)(45)(14)(13)(15)(25)(24)(35) \quad \text{(S.E. Rumsey)}.$$

2013:9. Let S be a set upon whose elements there is a binary operation $(x, y) \to xy$ which is associative (i.e. $x(yz) = (xy)z$). Suppose that there exists an element $e \in S$ for which $e^2 = e$ and that for each $a \in S$, there is at least one element b for which $ba = e$ and at most one element c for which $ac = e$. Prove that S is a group with this binary operation.

Solution 1. We show that e is the identity element of S and that each element of S has a two-sided inverse. Let t be an arbitrary element of S. There is an element $s \in S$ for which $st = e$. Hence $e = e^2 = s(tst)$. Since also $st = e$, we must have that $t = tst = te$. Therefore e is a right identity in S.

Suppose that $r \in S$ is such that $rs = e$. Then $et = rst = re = r$ (since e is a right identity). Therefore $ets = rs = e$. Since $ee = e$ as well, $e = ts$ and so $et = tst = t$.

Solution 2. [J. Love] Let $a \in S$. There exists b such that $ba = e$. Therefore, $b(ae) = (ba)e = e^2 = e$, so that $a = ae$ ($\forall a \in S$). We have that $e = ba = (be)a = b(ea)$ whereupon $a = ea$. Hence e is a two-sided identity.

Suppose that $ba = e$. Select $d \in S$ so that $db = e$. Then $d \doteq de = d(ba) = (db)a = ea = a$, so that a is a two-sided inverse of b.

[Problem **2013:9** is #4504 from the *American Mathematical Monthly* 59:8 (October, 1952), 554; 61:1 (January, 1954), 54–55.]

CHAPTER 10

Combinatorics and Finite Mathematics

2001:3.

    (a) Consider the infinite integer lattice in the plane (i.e., the set of points with integer coordinates) as a graph, with the edges being the lines of unit length connecting nearby points. What is the minimum number of colours that can be used to colour all the vertices and edges of this graph, so that

      (i) each pair of adjacent vertices gets two distinct colours;

      (ii) each pair of edges that meet at a vertex gets two distinct colours; and

      (iii) an edge is coloured differently than either of the two vertices at the ends?

    (b) Extend this result to lattices in real n-dimensional space.

Solution. Consider an n-dimensional lattice. Each vertex has $2n$ edges emanating from it (consisting of those points with exactly one coordinate differing by $+1$ or -1), so at least $2n + 1$ colours are needed. We create a colouring with this number of colours by building up one dimension at a time.

The integer points on the line and the edges between them can be coloured $1 - -(3) - -2 - -(1) - -3 - -(2) - -1$ and so on, where the edge colouring is in parentheses. Form a plane by stacking these lines unit distance apart, making sure that each vertex has a different coloured vertex above and below it; use colours 4 and 5 judiciously to colour the vertical edges. Now go to three dimensions; stack up planar lattices and struts unit distance apart, colouring each with the colours $1, 2, 3, 4, 5$, while making sure that vertically adjacent vertices have separate colours, and use the colours 6 and 7 for vertical struts. Continue on.

Comment. Here is a another colouring that will work for the plane: Let the colours be numbered $0, 1, 2, 3, 4$. Colour the point $(x, 0)$ with the colour

---

© Springer International Publishing Switzerland 2016   153
E.J. Barbeau, *University of Toronto Mathematics Competition*
(2001–2015), Problem Books in Mathematics,
DOI 10.1007/978-3-319-28106-3_10

$x \pmod 5$; colour the point $(0, y)$ with the colour $2y \pmod 5$; colour the points along each horizontal line parallel to the x-axis consecutively; colour the vertical edge whose lower vertex has colour $m \pmod 5$ with the colour $m + 1 \pmod 5$; colour the horizontal edge whose left vertex has the colour n $\pmod 5$ with the colour $n + 3 \pmod 5$.

2002:9. A sequence whose entries are 0 and 1 has the property that, if each 0 is replaced by 01 and each 1 by 001, then the sequence remains unchanged. Thus, it starts out as $010010101001\ldots$. What is the 2002th term of the sequence?

Solution. Let us define finite sequences as follows. Suppose that $S_1 = 0$. Then, for each $k \geq 2$, S_k is obtained by replacing each 0 in S_{k-1} by 01 and each 1 in S_{k-1} by 001. Thus,

$$S_1 = 0; \; S_2 = 01; \; S_3 = 01001; \; S_4 = 010010101001;$$

$$S_5 = 01001010100101001010010101001; \cdots$$

Each S_{k-1} is a prefix of S_k; in fact, it can be shown that, for each $k \geq 3$,

$$S_k = S_{k-1} * S_{k-2} * S_{k-1},$$

where $*$ indicates juxtaposition. The respective number of symbols in S_k for $k = 1, 2, 3, 4, 5, 6, 7, 8, 9, 10$ is equal to 1, 2, 5, 12, 29, 70, 169, 408, 985, 2378.

The 2002th entry in the given infinite sequence is equal to the 2002th entry in S_{10}, which is equal to the $(2002 - 985 - 408)$th $= (609)$th entry in S_9. This in turn is equal to the $(609 - 408 - 169)$th $= (32)$th entry in S_8, which is equal to the (32)th entry of S_6, or the third entry of S_3. Hence, the desired entry is 0.

2003:6. A set of n lightbulbs, each with an *on-off* switch, numbered $1, 2, \ldots, n$ are arranged in a line. All are initially off. Switch 1 can be operated at any time to turn its bulb on or off. Switch 2 can turn bulb 2 on or off if and only if bulb 1 is off; otherwise, it does not function. For $k \geq 3$, switch k can turn bulb k on or off if and only if bulb $k - 1$ is off and bulbs $1, 2, \ldots, k - 2$ are all on; otherwise it does not function.

    (a) Prove that there is an algorithm that will turn all of the bulbs on.
    (b) If x_n is the length of the shortest algorithm that will turn on all n bulbs when they are initially off, determine the largest prime divisor of $3x_n + 1$ when n is odd.

Solution. (a) Clearly $x_1 = 1$ and $x_2 = 2$. Let $n \geq 3$. The only way that bulb n can be turned on is for bulb $n - 1$ to be off and for bulbs $1, 2, \ldots, n - 2$ to be turned on. Once bulb n is turned on, then we need get bulb $n - 1$ turned on. The only way to do this is to turn off bulb $n - 2$; but for switch $n - 2$ to work, we need to have bulb $n - 3$ turned off. So before we can think about dealing with bulb $n - 1$, we need to get the first $n - 2$ bulbs turned off. Then we will be in the same situation as the outset with $n - 1$ rather than n bulbs. Thus the process has the following steps: (1) Turn on bulbs

$1, \ldots, n-2$; (2) Turn on bulb n; (3) Turn off bulbs $n-2, \ldots, 1$; (3) Turn on bulbs $1, 2, \ldots, n-1$. So if, for each positive integer k, y_k is the length of the shortest algorithm to turn off the first k bulbs after all are lit, then

$$x_n = x_{n-2} + 1 + y_{n-2} + x_{n-1} .$$

Since we can clearly find an algorithm for turning on or off bulb 1, an induction argument will provide an algorithm for turning any number of bulbs on or off.

We show that $x_n = y_n$ for $n = 1, 2, \ldots$. Suppose that we have an algorithm that turns all the bulbs on. We prove by induction that at each step we can legitimately reverse the whole sequence to get all the bulbs off again. Clearly, the first step is to turn either bulb 1 or bulb 2 on; since the switch is functioning, we can turn the bulb off again. Suppose that we can reverse the first $k - 1$ steps and are at the kth step. Then the switch that operates the bulb at that step is functioning and can restore us to the situation at the end of the $(k - 1)$th step. By the induction hypothesis, we can go back to having all the bulbs off. Hence, given the bulbs all on, we can reverse the steps of the algorithm to get the bulbs off again. A similar argument allows us to reverse the algorithm that turns the bulbs off. Thus, for each turning-on algorithm there is a turning-off algorithm of equal length, and vice versa. Thus $x_n = y_n$.

We have that $x_n = x_{n-1} + 2x_{n-2} + 1$ for $n \geq 3$. By, induction, we show that, for $m = 1, 2, \ldots$,

$$x_{2m} = 2x_{2m-1} \quad \text{and} \quad x_{2m+1} = 2x_{2m} + 1 = 4x_{2m-1} + 1.$$

This is true for $m = 1$. Suppose it is true for $m \geq 1$. Then

$$\begin{aligned} x_{2(m+1)} &= x_{2m+1} + 2x_{2m} + 1 = 2(x_{2m} + 1) + 4x_{2m-1} \\ &= 2(x_{2m} + 2x_{2m-1} + 1) = 2x_{2m+1}, \end{aligned}$$

and

$$\begin{aligned} x_{2(m+1)+1} &= x_{2(m+1)} + 2x_{2m+1} + 1 = 2x_{2m+1} + 4x_{2m} + 3 \\ &= 2(x_{2m+1} + 2x_{2m} + 1) + 1 = 2x_{2(m+1)} + 1. \end{aligned}$$

Hence, for $m \geq 1$,

$$3x_{2m+1} + 1 = 4(3x_{2m-1} + 1) = \cdots = 4^m(3x_1 + 1) = 4^{m+1} = 2^{2(m+1)}.$$

Thus, the largest prime divisor is 2.

2004:4. Let n be a positive integer exceeding 1. How many permutations $\{a_1, a_2, \ldots, a_n\}$ of $\{1, 2, \ldots, n\}$ are there which maximize the value of the sum

$$|a_2 - a_1| + |a_3 - a_2| + \cdots + |a_{i+1} - a_i| + \cdots + |a_n - a_{n-1}|$$

over all permutations? What is the value of this maximum sum?

The solution to this problem appears in Chap. 3.

2005:10. Let n be a positive integer exceeding 1. Prove that, if a graph with $2n+1$ vertices has at least $3n+1$ edges, then the graph contains a circuit (i.e., a closed non-self-intersecting chain of edges whose terminal point is its initial point) with an even number of edges. Prove that this statement does not hold if the number of edges is only $3n$.

Solution 1. If there are two vertices joined by two separate edges, then the two edges together constitute a chain with two edges. If there are two vertices joined by three distinct chains of edges, no pair sharing a vertex other than the endpoints, then the number of edges in two of the chains have the same parity, and these two chains together constitute a circuit with evenly many edges. Henceforth, we assume that neither of these situations occurs. We establish the general result by induction.

The result can be checked for $n = 2$. Then we have 5 vertices and at least 7 edges (but not more than 10). If we can find an even circuit among the seven edges, then *a fortiori* we can find one among more edges. Suppose that the edges $P - -Q$ and $P - -R$ and one other edge are missing. Then one of the circuits $P - -S - -R - -T - -P$, $Q - -S - -T - -R - -Q$, $P - -S - -T - -Q - -P$ will serve.

Suppose that the result holds for $2 \leq n \leq m - 1$. We may assume that we have a graph G that contains no instances where two separate edges join the same pair of vertices and no two vertices are connected by more than two distinct chains. Since $3n + 1 > 2n$, the graph is not a tree, and therefore must contain at least one circuit. Consider one of these circuits, L. If it has evenly many edges, the result holds. Suppose that it has oddly many edges, say $2k + 1$ with $k \geq 1$. Since any two vertices in the circuit are joined by at most two chains (the two chains that make up the circuit), there are exactly $2k + 1$ edges joining pairs of vertices in the circuit. Apart from the circuit, there are $(2m + 1) - (2k + 1) = 2(m - k)$ vertices and $(3m + 1) - (2k + 1) = 3(m - k) + k \geq 3(m - k) + 1$ edges.

We now create a new graph G', by coalescing all the vertices and edges of L into a single vertex v and retaining all the other edges and vertices of G. This graph G' contains $2(m - k) + 1$ vertices and at least $3(m - k) + 1$ edges, and so by the induction hypothesis, it contains a circuit M with an even number of edges. If this circuit does not contain v, then it is a circuit in the original graph G, which thus has a circuit with evenly many edges. If the circuit does contain v, it can be lifted to a chain in G joining two vertices of L by a chain of edges in G'. But these two vertices of L must coincide, for otherwise there would be three chains joining these vertices. Hence we get a circuit, *all* of whose edges lie in G'; this circuit has evenly many edges. The result now follows by induction.

Here is a counterexample with $3n$ edges. Consider $2n + 1$ vertices partitioned into a singleton and n pairs. Join each pair with an edge and join the singleton to each of the other vertices with a single edge to obtain a graph with $2n + 1$ vertices, $3n$ edges whose only circuits are triangles.

Solution 2. [J. Tsimerman] We can assume wolog that G has no double edges. For any graph H, let $k(H)$ be the number of circuits minus the number of components (two vertices being in the same component if and only if they are connected by a chain of edges). Let G_0 be the graph with $2n + 1$ vertices and no edges. Then $k(G_0) = -(2n + 1)$. Suppose that edges are added one at a time to obtain a succession G_i of graphs culminating in the graph $G = G_{3n+1}$ with $2n + 1$ vertices and $3n + 1$ edges. Since adding an edge either reduces the number of components (when it connects two vertices of separate components) or increases the number of circuits (when it connects two vertices in the same component), $k(G_{i+1}) \geq k(G_i) + 1$. Hence $k(G) \geq -(2n + 1) + (3n + 1) = n$. Thus, the number of circuits in G is at least equal to the number of components in G plus n, which is at least $n + 1$. Thus, G has at least $n + 1$ circuits.

Since $3(n + 1) > 3n + 1$, there must be two circuits that share an edge. If either circuit has an even number of edges, then the problem is solved. Otherwise, each has an odd number of edges. Suppose they share exactly one edge $C_0 - -D_0$. Indicating the two circuits by the numerals 1 and 2 and letting $- - k - -$ denote a path not containing a repeated edge in the circuit with *numeral* k, we can form a circuit with evenly many edges by cutting out the common edge:

$$C_0 - -1 - -D_0 - -2 - -C_0.$$

Since the deleted edge was counted twice in summing the edges of the two circuits, the new circuit has evenly many edges.

However, the two circuits could have more than one edge in common. We can still perform the above process to delete a single edge to get a single closed path that will have some edges traversed twice. This process of eliminating a single edge may present parity problems, so we present a method of dealing with two repeated edges $C_0 - -D_0$ and $C_1 - -D_1$ simultaneously. If there are two edges in common, the closed path can be a double circuit of one of two types.

The first type of closed path is:

$$C_0 - -D_0 - -1 - -C_1 - -D_1 - -1 - -C_0 - -D_0 - -2 - -D_1 - -C_1 - -2 - -C_0.$$

We can remove both edges (C_0, D_0) and (C_1, D_1) to get a closed path with evenly many edges:

$$C_0 - -2 - -C_1 - -1 - -D_0 - -2 - -D_1 - -1 - -C_0.$$

The second type of closed path is:

$$C_0 - -D_0 - -1 - -C_1 - -D_1 - -1 - -C_0 - -D_0 - -2 - -C_1 - -D_1 - -2 - -C_0.$$

Let there be p edges in the path $D_0 - -1 - -C_1$, q edges in $D_1 - -1 - -C_0$, r edges in $D_0 - -2 - -C_1$ and s edges in $D_1 - -2 - -C_0$. Note that $p + q$ and $r + s$ are both odd. If p and r have the opposite parity, then p and s have the same parity and $C_0 - -D_0 - -1 - -C_1 - -D_1 - -2 - -C_0$ is a closed path with evenly many edges. If p and r have the same parity, then $D_0 - -1 - -C_1 - -2 - -D_0$ is a closed path with evenly many edges.

Thus, we can take our two circuits and by bifurcating at a common edge obtain a single closed path with some finite number of edges traversed twice. By the foregoing process, we can eliminate the repeated edges one or two at a time until we end up with the required type of circuit.

A counterexample can be obtained by taking a graph with vertices A_1, ..., A_n, B_0, B_1, ..., B_n, with edges joining the vertex pairs (A_i, B_{i-1}), (A_i, B_i) and (B_{i-1}, B_i) for $1 \leq i \leq n$.

2006:1.

(a) Suppose that a 6×6 square grid of unit squares (chessboard) is tiled by 1×2 rectangles (dominoes). Prove that it can be decomposed into two rectangles, tiled by disjoint subsets of the dominoes.

(b) Is the same thing true for an 8×8 array?

Solution. (a) There are 18 dominoes and 10 interior lines in the grid. Suppose, if possible, that the decomposition does not occur for some disposition of dominoes. Then each of the lines must be straddled by at least one domino. We argue that, in fact, at least two dominoes must straddle each line. Since no domino can straddle more than one line, this would require 20 dominoes and so yield a contradiction.

Each interior line has six segments. For a line next to the side of the square grid, an adjacent domino between it and the side must either cross one segment or be adjacent to two segments. Since the number of segments is even, evenly many dominoes (at least two) must cross a segment. For the next line in, an adjacent domino must be adjacent to two segments, be adjacent to one segment and cross the previous line, or cross one segment. Since the number of dominoes straddling the previous line is even, there must be evenly many that cross the segment. In this way, we can work our way from one line to the next.

(b) Figure 10.1 shows a possible arrangement of the dominoes.

2006:5. Suppose that you have a 3×3 grid of squares. A *line* is a set of three squares in the same row, the same column or the same diagonal; thus, there are eight lines.

Two players A and B play a game. They take alternate turns, A putting a 0 in any unoccupied square of the grid and B putting a 1. The first player is A, and the game cannot go on for more than nine moves. (The play is similar to noughts and crosses, or tic-tac-toe.) A move is *legitimate* if it does not result in two lines of squares being filled in with different sums. The winner is the last player to make a legitimate move.

(For example, if there are three 0s down the diagonal, then B can place a 1 in any square provided it completes no other line, for then the sum would differ from the diagonal sum. If there are two zeros at the top of the main

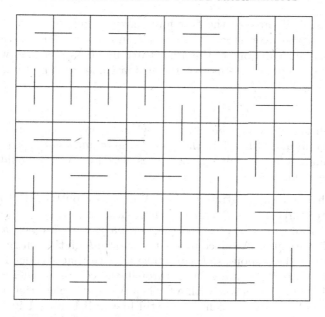

Fig. 10.1. A nondecomposable tiling of dominoes

diagonal and two ones at the left of the bottom line, then the lower right square cannot be filled by either player, as it would result in two lines with different sums.)

(a) What is the maximum number of legitimate moves possible in a game?

(b) What is the minimum number of legitimate moves possible in a game that would not leave a legitimate move available for the next player?

(c) Which player has a winning strategy? Explain.

Solution. (a) A game cannot continue to nine moves. Otherwise, each line sum must be three times the value on the centre square of the grid (why?) and so must be 0 or 3. But some line must contain both zeros and ones, yielding a contradiction. [An alternative argument is that, if the array is filled, not all the rows can have the same numbers of 0s and 1s, and therefore cannot have the same sums.] However, an eight-move grid is possible, in which one player selects the corner squares and the other the squares in the middle of the edges.

(b) Consider any game after four moves have occurred and it is A's turn to play a zero. Suppose, first of all, that no lines have been filled with numbers. The only way an inaccessible square can occur is if it is the intersection of two lines each having the other two squares filled in. This can happen in at most one way. So A would have at least four possible squares to fill in. On the other hand, if three of the first four moves complete a line, the fourth number can bar at most three squares for A in the three lines determined by the fourth number and one of the other three numbers. Thus, A would have at least two possible positions to fill. Thus, a game must go to at least five moves.

A five-move game can be obtained when A has placed three 0's down the left column and B has 1 in the centre square and another square of the middle column. Each remaining position is closed to B as it would complete a line whose sum is not 0.

(c) A has a winning strategy. Let A begin by putting 0 in the centre square. After three moves, A can guarantee either a row, column or diagonal with two 0's and one 1. Suppose that we have 0, 0, 1 down a diagonal. Wherever B puts his 1, A can complete the line through the centre containing it. For each of B's responses, A can make a move that stymies B. Suppose that we have 1, 0, 0 down a column through the centre parallel to a side of the grid. If B puts 1 in a row with a 0, then A can complete that row. B has one possible move, after which A can still move. If B puts 1 in a row with another 1, A puts a 0 at the other end of the diagonal. B has only one move, to which A can respond.

2007:1. A $m \times n$ rectangular array of distinct real numbers has the property that the numbers in each row increase from left to right. The entries in each column, individually, are rearranged so that the numbers in each column increase from top to bottom. Prove that in the final array, the numbers in each row will increase from left to right.

Solution 1. We prove this by induction. Note that the permutation yielding the final arrangement of each column is uniquely determined, so that if we can perform a sequence of transposition (switches) resulting in the entries increasing from top to bottom, the composite of these transpositions is the required permutation.

We can arrange the rows so that the first column is increasing from top to bottom; all the rows will still be increasing from left to right. Suppose that we have performed a sequence of rearrangements of sets of columns so that (a) each row is increasing, and (b) the first k columns are increasing for $1 \leq k \leq n - 1$. Let the entries in the kth column be $a_1, a_2, \ldots, a_m$ and in the $(k+1)$th column be $b_1, b_2, \ldots, b_m$. We have that $a_1 \leq a_2 \leq a_3 \leq \cdots \leq a_m$ and $a_i \leq b_i$ for $1 \leq i \leq m$.

Suppose that b_r is the minimum of all the b_i ($1 \leq i \leq m$). We interchange the elements in the first and rth rows of the jth column for $k + 1 \leq j \leq n$. Since $a_1 \leq a_r \leq b_r$ and $a_r \leq b_r \leq b_1$, the first and rth rows are still increasing.

Let b_s be the minimum of all the b_i except for b_r; interchange the elements in the second and sth rows of the jth column for $k + 1 \leq j \leq n$. Since $a_2 \leq a_s \leq b_s$ and $a_s \leq b_s \leq b_2$, the new second and sth rows are increasing.

Observe that $b_r \leq b_s \leq b_i$ for $i \neq r, s$, so that the $(k+1)$th column is increasing down to the third entry.

We can continue in this way, moving the third smallest b_i to the third row, and so on, ending up with changing the order of the columns from the $(k+1)$th to the nth and keeping the rows increasing. The result follows by induction on k.

Solution 2. [M. Cai] Let $a_{i,j}$ be the (i,j)th entry in the array before the rearrangement and $b_{i,j}$ the (i,j)th entry after the rearrangement. Then

$$b_{1,j} \leq b_{2,j} \leq \cdots \leq b_{m,j}$$

for $1 \leq j \leq n$. We need to show that, for each i with $1 \leq i \leq m$,

$$b_{i,1} \leq b_{i,2} \leq \cdots \leq b_{i,n}.$$

Let $1 \leq j \leq n-1$. For $1 \leq r \leq m$, we have that

$$b_{m,j+1} \geq a_{r,j+1} \quad \text{and} \quad a_{r,j} \leq a_{r,j+1}.$$

There exist s with $1 \leq s \leq m$ for which $b_{m,j} = a_{s,j}$. Hence, for $1 \leq j \leq n-1$,

$$b_{m,j} = a_{s,j} \leq a_{s,j+1} \leq b_{m,j+1}.$$

Suppose, as an induction hypothesis, it has been established that $b_{i,j} \leq b_{i,j+1}$ for $2 \leq k+1 \leq i \leq m$ and $1 \leq j \leq n-1$. Then $b_{k,j} = a_{t,j} \leq a_{t,j+1}$ for some t with $1 \leq t \leq m$. Since the set of numbers $\{b_{i,j+1} : k \leq i \leq m\}$ is the set of the largest $m-k+1$ numbers of the set $\{a_{i,j+1} : 1 \leq i \leq m\}$, we must have that $a_{t,j+1} \leq b_{t,j+1}$. The result follows.

Solution 3. There is nothing to prove when $n = 1$. Suppose that $n \geq 2$, that a given column, not the last, is $A = (a_1, a_2, \ldots, a_m)^t$ to begin with and $B = (b_1, b_2, \ldots, b_m)^t$ after the rearrangement, so that $b_1 \leq b_2 \leq \cdots \leq b_m$. Let the elements in a column to the right of this be $C = (c_1, c_2, \ldots, c_m)^t$ to begin with and $D = (d_1, d_2, \ldots, d_m)^t$ after the rearrangement. We have that $d_1 \leq d_2 \leq \cdots \leq d_m$ and $a_i \leq c_i$ for $1 \leq i \leq m$. We need to show that $b_k \leq d_k$ for $1 \leq k \leq m$.

Since b_k is the kth largest number in the given column A to the left, there are $m-k+1$ elements in that column not less than it. Hence there are at least $m-k+1$ elements in the column C to the right that are not less than b_k, and these elements include $d_k, d_{k+1}, \ldots, d_m$. Hence $b_k \leq d_k$.

2007:3. Prove that the set $\{1, 2, \ldots, n\}$ can be partitioned into k subsets with the same sum if and only if k divides $\frac{1}{2}n(n+1)$ and $n \geq 2k-1$.

Solution. The necessity of the conditions follows from the fact that the sum of all the numbers in the set is $\frac{1}{2}n(n+1) = ks$, where s is the sum of the numbers in each subset, and there is at most one subset with a single element; hence the number k of subsets is at most $\frac{1}{2}(n-1)+1$. (*Alternatively:* Since n must lie in one of the subsets, the sum of each subset is at least n, and so $kn \leq \frac{1}{2}n(n+1)$. Therefore $n \geq 2k-1$.)

On the other hand, let k, s and n be positive integers for which $ks = \frac{1}{2}n(n+1)$ and $2k - 1 \leq n$. Then $ks \geq \frac{1}{2}(2k-1)(2k)$ from which $s \geq 2k - 1$, $k \leq \frac{1}{2}(s+1)$ and $\frac{1}{2}n(n+1) = ks \leq \frac{1}{2}(s+1)s$. Thus both n and k do not exceed s.

For $k = 1$ and each value of s, the conditions holds with $n = s$, which corresponds to the partition given by the single set $(1, 2, \ldots, s) = (1, 2, \ldots, n)$. The only triples (k, n, s) satisfying the conditions with $s \leq 3$ are $(k, n, s) = (1, 1, 1)$, $(1, 2, 3)$ and $(2, 3, 3)$, corresponding respectively to the partitions $(\{1\})$, $(\{1, 2\})$ and $(\{1, 2\}, \{3\})$.

We establish the result by double induction on s and k. Suppose that the result has been established for all possible subset sums up to $s - 1 \geq 3$, as well as for s and possible number of subsets in the partition up to $k - 1 \geq 1$. Consider the case where the conditions are satisfied for (k, n, s), and we have to find a suitable partition. Let $n = 2k - 1$, so that $s = n(n+1)/(2k) = 2k - 1$. Then a partition of the required type consists of the singleton $\{2k - 1\}$ and the $k - 1$ pairs $\{i, 2k - 1 - i\}$ with $1 \leq i \leq k - 1$. If $n = 2k$, then $s = n(n+1)/(2k) = 2k + 1$ and we can use the k pairs $\{i, 2k + 1 - i\}$ with $1 \leq i \leq k$. We use induction on n to establish the result for $n > 2k$.

Suppose that $2k < n < 4k - 1$ and that $s = n(n+1)/(2k)$ is an integer. We have that $2 < n/k < (n+1)/k < 4$ so that

$$n + 1 = \frac{2ks}{n} < s = 2n\left(\frac{n+1}{4k}\right) < 2n.$$

Suppose that $n' = s - n - 1$, so that $0 < n' < n - 1$.

First, let s be odd. The sum of the numbers from 1 to $s - n - 1$ inclusive is equal to

$$\frac{(s-n)(s-n-1)}{2} = \frac{s^2 - (2n+1)s}{2} + \frac{n(n+1)}{2} = s\left[\frac{s-1}{2} - n + k\right].$$

Let $k' = \frac{1}{2}(s - 1) - n + k$; note that $k' < k$ and that

$$n' - (2k' - 1) = (s - n - 1) - (s - 2 - 2n + 2k) = n - 2k + 1 > 0,$$

so that $n' > 2k' - 1$. By the induction hypothesis, we can partition $\{1, 2, \ldots, s - n - 1\}$ into k' subsets, each of whose sum is s. Let $r = \frac{1}{2}(s - 1)$, and adjoin the $n - r = k - k'$ doubletons $\{s - n, n\}, \{s - n + 1, n - 1\}, \ldots, \{r, r + 1\}$ to get the desired partition of $\{1, 2, \ldots, n\}$ into k subsets with the same sum s.

Now let s be even. Then

$$\frac{(s-n)(s-n-1)}{2} = \frac{s}{2}[s - 1 - 2n + 2k].$$

Let $k' = 2k - 2n + s - 1 = (n - 2k)(n + 1 - 2k)/(2k)$ and $s' = s/2$. Note that k' need not be less than k, but in any case the new purported subset sum s' is less than s. Since $s < 2(n - 1)$, $2s' = s > 2(s - n - 1) = 2n'$. By the induction hypothesis, we can partition the set $\{1, 2, \ldots, s - n - 1\}$ into k' subsets whose sums are all s'. Since k' is odd and since $s' > s - n - 1$, we can augment this family of subsets by the singleton $\{s'\}$ to get evenly many subsets with sum s'. Pair them off to get $\frac{1}{2}(k'+1) = k - n + s'$ subsets with sum s, and further augment the family with the $n - s'$ doubletons $\{s - n, n\}$, $\{s - n + 1, n - 1\} \ldots, \{s' - 1, s' + 1\}$ to get a partition of $\{1, 2, \ldots, n\}$ into k sets with sum s.

Finally, suppose that $n \geq 4k - 1$. Let $n' = n - 2k$, $k' = k$. Then, if $s = n(n+1)/(2k)$, then the sum $s' = n'(n'+1)/2k$ is an integer equal to

$$\frac{(n - 2k)(n - 2k + 1)}{2k} = \frac{n(n+1)}{2k} - (2n + 1) + 2k,$$

that is not less than $2k - 1 > 0$. Note that $n' = n - 2k \geq 2k - 1 = 2k' - 1$ and that

$$s - s' = \frac{(n - n')(n + n' + 1)}{2k} = 2n - 2k + 1.$$

Determine a partition of $\{1, 2, \ldots, n - 2k\}$ into k subsets with sum s', and adjoin to these subsets the k doubletons

$$\{n - 2k + 1, n\}, \{n - 2k + 2, n - 1\}, \ldots, \{n - k, n - k + 1\}.$$

The whole result is now established.

Comment. The case $2k < n < 4k - 1$ with k a divisor of $\frac{1}{2}n(n + 1)$ is not realizable for $k = 2$ and $k = 4$. For $k \equiv 1 \pmod 4$ and s odd, a possible value of n is $3k - 1$, and following the notation in the solution, we get the possibilities

$$(k, n, s) = \left(k, 3k - 1, \frac{3(3k - 1)}{2}\right)$$

$$(k', n', s') = \left(\frac{k - 1}{4}, \frac{3(k - 1)}{2}, \frac{3(3k - 1)}{2}\right).$$

For $k \equiv 1 \pmod 4$ and s even, we have the possibilities

$$(k, n, s) = \left(k, 3k, \frac{3(3k + 1)}{2}\right)$$

$$(k', n', s') = \left(\frac{k + 1}{2}, \frac{3(k + 1)}{2}, \frac{3(3k + 1)}{4}\right).$$

For $k \equiv 3 \pmod 4$ and s odd, we have the possibilities

$$(k, n, s) = \left(k, 3k, \frac{3(3k + 1)}{2}\right)$$

$$(k', n', s') = \left(\frac{k + 1}{4}, \frac{3(k + 1)}{2}, \frac{3(3k + 1)}{2}\right).$$

For $k \equiv 3 \pmod 4$ and s even, we have the possibilities

$$(k, n, s) = \left(k, 3k - 1, \frac{3(3k - 1)}{2}\right)$$

$$(k', n', s') = \left(\frac{k-1}{2}, \frac{3(k-1)}{2}, \frac{3(3k-1)}{4} \right).$$

This exhausts the possibilities for $k \leq 5$. For $k = 6$, there is one possibility:

$$(k, n, s) = (6, 15, 20); (k', n', s') = (1, 4, 10)$$

with the partition

$$\{1, 2, 3, 4, 10\}, \{5, 15\}, \{6, 14\}, \{7, 13\}, \{8, 12\}, \{9, 11\}.$$

In dealing with the case of even s, it can indeed occur that $k' > k$ (which is why we have to extend the induction to s as well as k). Applying the reduction to $(k, n, s) = (14, 48, 84)$ leads to $(k', n', s') = (15, 35, 42)$ which in turn reduces to $(k'', n'', s'') = (1, 6, 21)$. The partition that corresponds to $(14, 48, 84)$ turns out to be

$$\{1, 2, 3, 4, 5, 6, 7, 21, 42\}, \{7, 8, 34, 35\}, \{9, 10, 32, 33\}, \{11, 12, 30, 31\},$$

$$\{13, 14, 28, 29\}, \{15, 16, 26, 27\}, \{17, 18, 24, 25\}, \{19, 20, 22, 23\}, \{36, 48\}, \{37, 47\},$$

$$\{38, 46\}, \{39, 45\}, \{40, 44\}, \{41, 43\}.$$

2008:6. 2008 circular coins, possibly of different diameters, are placed on the surface of a flat table in such a way that no coin is on top of another coin. What is the largest number of points at which two of the coins could be touching?

Solution. Suppose that there are n coins where $n \geq 3$. We show that the answer is $3n - 6$. When $n = 2008$, this number is equal to 6018. We can achieve this as follows. Start with three mutually touching coins. Insert a coin of suitable diameter in the middle so that it touches all three coins. Continue adding coins, each in the middle of three mutually touching coins that have already been placed.

To show that this is an upper bound, construct a graph whose vertices are the centres of the coins, with two vertices connected by an edge if and only if the their two coins touch. This is a planar graph with n vertices, with the number of edges equal to the number of points where two coins touch. We use the following result from graph theory: *Any planar graph with V vertices, E edges and F faces (including the unbounded face) has at most $3V - 6$ edges.*

Since each face has at least three edges, and each edge belongs to two faces, we must have $2E \geq 3F$. Hence, using Euler's equation,

$$2 = V - E + F \leq V - E + \frac{2}{3}E = V - \frac{1}{3}E,$$

whence $E \leq 3V - 6$.

[Problem **2008:6** was contributed by David Arthur.]

2008:10. A point is chosen at random (with the uniform distribution) on each side of a unit square. What is the probability that the four points are the vertices of a quadrilateral with area exceeding $\frac{1}{2}$?

The solution to this problem appears in Chap. 8.

2009:10. Suppose that a path on an $m \times n$ grid consisting of the lattice points $\{(x, y) : 1 \leq x \leq m, 1 \leq y \leq n\}$ (x and y both integers) consisting of $mn - 1$ unit segments begins at the point $(1, 1)$, passes through each point of the grid exactly once, does not intersect itself and finishes at the point (m, n). Show that the path partitions the rectangle bounded by the lines $x = 1$, $x = m$, $y = 1$, $y = n$ into two subsets of equal area, the first consisting of regions opening to the left or up, and the second consisting of regions opening to the right or down.

Solution. Embed the grid into a rectangle composed of mn unit square cells, each with a grid point as centre. Extend the path horizontally to the left and right end of this rectangle from the respective points $(1, 1)$ and (m, n). It is equivalent to show that this rectangle is decomposed by this extended path into two subsets of equal area. As the path proceeds, at any grid point, it either turns left through a right angle, or turns right through a right angle or proceeds straight ahead. Since it ends up going in the same direction as it started, it makes an equal number of left and right turns. The path keeps the region opening left or up on its left and the region opening right or down on its right. At each left turn, the path splits the area of the cell it is in into two subsets of areas $1/4$ on the left and $3/4$ on the right; at each right turn, it splits the cell in the opposite way. If the path goes straight ahead, it splits the cell area in half. It follows from this that the area on the left side of the path equals the area on the right side of the path, and the result holds.

2010:2. Let $u_0 = 1$, $u_1 = 2$ and $u_{n+1} = 2u_n + u_{n-1}$ for $n \geq 1$. Prove that, for every nonnegative integer n,

$$u_n = \sum \left\{ \frac{(i+j+k)!}{i!j!k!} : i, j, k \geq 0, i + j + 2k = n \right\}.$$

Solution 1. Suppose that we have a supply of white and of blue coaches, each of length 1, and of red coaches, each of length 2; the coaches of each colour are indistinguishable. Let v_n be the number of trains of total length n that can be made up of red, white and blue coaches of total length n. Then $v_0 = 1$, $v_1 = 2$ and $v_2 = 5$ (R, WW, WB, BW, BB). In general, for $n \geq 1$, we can get a train of length $n + 1$ by appending either a white or a blue coach to a train of length n or a red coach to a train of length $n - 1$, so that $v_{n+1} = 2v_n + v_{n-1}$. Therefore $v_n = u_n$ for $n \geq 0$.

We can count v_n in another way. Suppose that the train consists of i white coaches, j blue coaches and k red coaches, so that $i + j + 2k = n$. There are $(i+j+k)!$ ways of arranging the coaches in order; any permutation of the i white coaches among themselves, the j blue coaches among themselves and k red coaches among themselves does not change the train. Therefore

$$u_n = \sum \left\{ \frac{(i+j+k)!}{i!j!k!} : i,j,k \geq 0, i+j+2k = n \right\}.$$

Solution 2. Let $f(t) = \sum_{n=0}^{\infty} u_n t^n$. Then

$$\begin{aligned}
f(t) &= u_0 + u_1 t + (2u_1 + u_0)t^2 + (2u_2 + u_1)t^3 + \cdots \\
&= u_0 + u_1 t + 2t(f(t) - u_0) + t^2 f(t) \\
&= u_0 + (u_1 - 2u_0)t + (2t + t^2)f(t) \\
&= 1 + (2t + t^2)f(t),
\end{aligned}$$

whence

$$f(t) = \frac{1}{1 - 2t - t^2} = \frac{1}{1 - t - t - t^2} = \sum_{n=0}^{\infty}(t + t + t^2)^n$$

$$= \sum_{n=0}^{\infty} \sum \left\{ \frac{(i+j+k)!}{i!j!k!} t^{i+j+2k} : i,j,k \geq 0, i+j+2k = n \right\}.$$

For a given exponent $m \geq 0$, we need to determine the coefficient of t^m in the right sum. The values of k range between 0 and $\lfloor m/2 \rfloor$, and for each we get a term in t^m when $i + j + k = m - k$, so the coefficient of t^m is

$$\sum_{k=0}^{\lfloor m/2 \rfloor} \left\{ \frac{(i+j+k)!}{i!j!k!}; i,j \geq 0, i+j+k = m - k \right\}$$

$$\sum \left\{ \frac{(i+j+k)!}{i!j!k!}; i,j,k \geq 0, i+j+k = m - k \right\}$$

Replacing m by n, we find that

$$f(t) = \left[\sum \left\{ \frac{(i+j+k)!}{i!j!k!} : i,j,k \geq 0, i+j+2k = n \right\} \right],$$

from which the desired result follows.

Solution 3. Let w_n be the sum in the problem. It is straightforward to check that $u_0 = w_0$ and $u_1 = w_1$. We show that, for $n \geq 1$, $w_{n+1} = 2w_n + w_{n-1}$ from which it follows by induction that $u_n = w_n$ for each n. By convention, let $(-1)! = \infty$. Then, for $i,j,k \geq 0$ and $i + j + 2k = n + 1$, we have that

$$\begin{aligned}
\frac{(i+j+k)!}{i!j!k!} &= \frac{(i+j+k)(i+j+k-1)!}{i!j!k!} \\
&= \frac{(i+j+k-1)!}{(i-1)!j!k!} + \frac{(i+j+k-1)!}{i!(j-1)!k!} + \frac{(i+j+k-1)!}{i!j!(k-1)!},
\end{aligned}$$

whence

$$
\begin{aligned}
w_{n+1} &= \sum \left\{ \frac{(i+j+k-1)!}{(i-1)!j!k!} : i,j,k \geq 0, (i-1)+j+2k = n \right\} \\
&+ \sum \left\{ \frac{(i+j+k-1)!}{i!(j-1)!k!} : i,j,k \geq 0, i+(j-1)+2k = n \right\} \\
&+ \sum \left\{ \frac{(i+j+k-1)!}{i!j!(k-1)!} : i,j,k \geq 0, i+j+2(k-1) = n-1 \right\} \\
&= w_n + w_n + w_{n-1} = 2w_n + w_{n-1}
\end{aligned}
$$

as desired.

[Problem **2010:2** is #10663 in *American Mathematical Monthly* 105:5 (May, 1998), 464; 107:4 (April, 2000), 370–371.]

2010:4. The plane is partitioned into n regions by three families of parallel lines. What is the least number of lines to ensure that $n \geq 2010$?

Solution. Suppose that there are x, y and z lines in the three families. Assume that no point is common to three distinct lines. The $x + y$ lines of the first two families partition the plane into $(x+1)(y+1)$ regions. Let λ be one of the lines of the third family. It is cut into $x+y+1$ parts by the lines in the first two families, so the number of regions is increased by $x+y+1$. Since this happens z times, the number of regions that the plane is partitioned into by the three families of lines is

$$
n = (x+1)(y+1) + z(x+y+1) = (x+y+z) + (xy+yz+zx) + 1.
$$

Let $u = x+y+z$ and $v = xy+yz+zx$. Then (by the Cauchy-Schwarz Inequality for example), $v \leq x^2+y^2+z^2$, so that $u^2 = x^2+y^2+z^2+2v \geq 3v$. Therefore, $n \leq u + \frac{1}{3}u^2 + 1$. This is a value less than 2003 when $u = 76$. However, when $(x,y,z) = (26,26,25)$, then $u = 77$, $v = 1976$ and $n = 2044$. Therefore, we need at least 77 lines, but a suitably chosen set of 77 lines will suffice.

2011:3. Suppose that S is a set of n nonzero real numbers such that exactly p of them are positive and exactly q are negative. Determine all the pairs (n,p) such that exactly half of the threefold products abc of distinct elements a,b,c of S are positive.

Solution 1. A threefold product is positive if and only if all factors are positive or exactly one factor is positive; it is negative if and only if all factors are negative or exactly one factor is negative. This leads to the equation

$$
\binom{p}{3} + p\binom{q}{2} = \binom{q}{3} + q\binom{p}{2}.
$$

This simplifies to

$$
(p-q)[(p-q)^2 - 3(p+q) + 2] = 0.
$$

We have that $n = p + q$; let $m = |p - q|$. Then the condition is satisfied if and only if either $p = q$ and $n = 2p$ is even, or $n = \frac{1}{3}(m^2 + 2)$, m is not divisible by 3, and p and q are equal to $(m + 1)(m + 2)/6$ and $(m - 1)(m - 2)/6$ in some order.

Solution 2. Since half of the total number of products are to be positive, we obtain the condition

$$\binom{p}{3} + p\binom{n-p}{2} = \frac{1}{2}\binom{n}{3}.$$

This reduces to

$$2p(p-1)(p-2) + 6p(n-p)(n-p-1) = n(n-1)(n-2),$$

and ultimately to

$$0 = (8p^3 - 12p^2 n + 6pn^2 - n^3) + (-6pn + 3n^2) + (4p - 2n)$$
$$= (2p - n)[(2p - n)^2 - 3n + 2].$$

Therefore, we must have either that $2p = n$ or $n = \frac{1}{3}(m^2 + 2)$ where $m = |2p - n|$. The argument can be completed as in Solution 1.

Solution 3. There is no loss of generality in assuming that all the positive numbers are $+1$ and the negative numbers are -1. Then, when exactly half the threefold products are positive, the sum of all the threefold products vanishes. Let $m = p - q$, so that m is the sum of the numbers in S. Then

$$m^3 = \sum\{a^3 : a \in S\} + 3\sum\{a^2 b : a, b \in S\} + 6\sum\{abc : a, b, c \in S\}$$
$$= m + 3(n-1)m + 0$$

whence either $m = 0$ or $m^2 = 3n - 2$. This leads to the same conclusion as in Solution 1.

[Problem **2011:3** is #10683 in the *American Mathematical Monthly* 105:8 (October, 1998), 768; 107:8 (October, 2000), 754.]

2011:6. [The problem posed on the competition is not available for publication, as it appeared as Enigma problem 1610 in the August 25, 2010 issue of the *New Scientist*. It concerned the determination of the final score of a three game badminton match where the score of each competitor was in arithmetic progression. The full statement of the problem is available on the internet.]

The solution to this problem appears in Chap. 2.

2011:7. Suppose that there are 2011 students in a school and that each student has a certain number of friends among his schoolmates. It is assumed that if A is a friend of B, then B is a friend of A, and also that there may exist certain pairs that are not friends. Prove that there is a nonvoid subset S of these students for which every student in the school has an even number of friends in S.

The solution to this problem appears in Chap. 7.

2011:10. Suppose that p is an odd prime. Determine the number of subsets S contained in $\{1, 2, \ldots, 2p - 1, 2p\}$ for which (a) S has exactly p elements, and (b) the sum of the elements of S is a multiple of p. [**3**]

The solution to this problem appears in Chap. 2.

2012:9. In a round-robin tournament of $n \geq 2$ teams, each pair of teams plays exactly one game that results in a win for one team and a loss for the other (there are no ties).

(a) Prove that the teams can be labelled $t_1, t_2, \ldots, t_n$, so that, for each i with $1 \leq i \leq n - 1$, team t_i beats t_{i+1}.

(b) Suppose that a team t has the property that, for each other team u, one can find a chain $u_1, u_2, \ldots, u_m$ of (possibly zero) distinct teams for which t beats u_1, u_i beats u_{i+1} for $1 \leq i \leq m - 1$ and u_m beats u. Prove that *all* of the n teams can be ordered as in (a) so that $t = t_1$ and each t_i beats t_{i+1} for $1 \leq i \leq n - 1$.

(c) Let T denote the set of teams that can be labelled as t_1 in an ordering of teams as in (a). Prove that, in any ordering of teams as in (a), all the teams in T occur before all the teams that are not in T.

Preliminaries. Let $a > b$ denote that team a beats team b. The result is clear when $n = 2$. Assume that the result holds for all numbers of teams up to $n - 1 \geq 2$.

(a) *Solution 1.* By the induction hypothesis, we can order $n - 1$ of the teams so that $t_1 > t_2 > \cdots > t_{n-1}$. Suppose that t is the nth team not included in this list. If t either beats t_1 or is beaten by t_{n-1} then it can be appended at one end of the list. Suppose therefore that $t_1 > t > t_{n-1}$. Let i be the largest index for which $t_i > t$. Then $i < n - 1$ and we can insert t between t_i and t_{i+1}.

Solution 2. Let t be any team, A be the set of teams that beat t and B be the set of teams that are beaten by t. We use induction. If A is nonvoid, then the teams in A can be ordered as in (a) so that each team beats the next; similarly, if B is nonvoid, its teams can be ordered in the same way. The orderings in A and B with t interpolated give the required ordering of all the teams.

(b) *Solution 1.* Suppose that $t > u_1 > u_2 > \cdots > u_r$ is a list of teams, each beating the next, of maximum length. Suppose, if possible, that a team v is not included in the list. Then $v > u_r$. If there is a team u_i with $u_i > v$, then by taking such a team with the maximum value of i, we can insert v between u_i and u_{i+1}, contradicting the maximality of the list. Therefore $v > u_i$ for each i.

By the hypothesis on t, we can find teams $v_1, v_2, \ldots, v_s$ for which $t > v_1 > v_2 > v_3 > \cdots > v_s = v$. Consider team v_{s-1}. This team is not one of the u_i since it beats $v_s = v$. Therefore, arguing as for v, we find that

$v_{s-1} > u_r$. If, for some i, $u_i > v_{s-1}$, then we could insert $> v_{s-1} > v_s >$ between u_i and u_{i+1} in the list, contradicting maximality. Hence $v_{s-1} > u_i$ for each i. We can continue on in this way to argue in turn that $\dot{v}_{s-2}, \ldots,$ v_1 are all distinct from the u_i and beat each of the u_i.

But then, we could form a chain

$$t > v_1 > v_2 > \cdots > v_s > u_1 > \cdots > u_r = u$$

again contradicting the maximality. Hence any maximum chain contains all the teams.

Solution 2. [J. Zung] We can consider the set T of teams as a directed graph with root t where each pair of vertices is connected by a directed edge $a \to b$, corresponding to team a beating team b. Let T_1 be a minimum spanning tree of T with root t, (i.e., T and T_1 have the same vertices, T_1 is connected, but T_1 has no loops, and there is a path from t to any other vertex). If every vertex in T_1 has at most one exiting edge, then T_1 is a path of all the vertices. Otherwise, there are vertices u, v, w for which $u \to v$, $u \to w$. Suppose, wolog, in T, that $v \to w$. We convert T_1 to a new tree T_2 with one fewer vertex having more than one exiting edge by removing the edge $u \to w$ and inserting the edge $v \to w$.

We can continue the same process to get a succession of spanning trees from which the sum over all vertices of the number of edges required to go from t to those nodes increases. The process must terminate with a spanning path.

(c) *Solution.* First note that any team x that beats a team t in T must itself belong to T. For let an ordering of the all the teams be given and append x at the beginning. We must then have an ordering

$$x > t > \cdots > x > \cdots$$

in which x appears twice and every other team once. It follows from this that any team is at the end of a chain that begins with one of the x and consists of distinct teams. Thus, by (b), any team beating a member of T must lie in T. The desired result follows from this.

2015:2. Given $2n$ distinct points in space, the sum S of the lengths of all the segments joining pairs of them is calculated. Then n of the points are removed along with all the segments having at least one endpoint from among them. Prove that the sum of the lengths of all the remaining segments is less than $\frac{1}{2}S$.

The solution to this problem appears in Chap. 8.

CHAPTER 11

Number Theory

2002:3. In how many ways can the rational $2002/2001$ be written as the product of two rationals of the form $(n+1)/n$, where n is a positive integer?

Solution. We begin by proving a more general result. Let m be a positive integer, and denote by $\tau(m)$ and $\tau(m+1)$, the number of positive divisors of m and $m+1$ respectively. Suppose that

$$\frac{m+1}{m} = \frac{p+1}{p} \cdot \frac{q+1}{q},$$

where p and q are positive integers exceeding m. Then $(m+1)pq = m(p+1)(q+1)$, which reduces to $(p-m)(q-m) = m(m+1)$. It follows that $p = m+u$ and $q = m+v$, where $uv = m(m+1)$. Hence, every representation of $(m+1)/m$ corresponds to a factorization of $m(m+1)$.

On the other hand, observe that, if $uv = m(m+1)$, then

$$
\begin{aligned}
\frac{m+u+1}{m+u} \cdot \frac{m+v+1}{m+v} &= \frac{m^2 + m(u+v+2) + uv + (u+v) + 1}{m^2 + m(u+v) + uv} \\
&= \frac{m^2 + (m+1)(u+v) + m(m+1) + 2m + 1}{m^2 + m(u+v) + m(m+1)} \\
&= \frac{(m+1)^2 + (m+1)(u+v) + m(m+1)}{m^2 + m(u+v) + m(m+1)} \\
&= \frac{(m+1)[(m+1) + (u+v) + m]}{m[m + (u+v) + m + 1]} = \frac{m+1}{m}.
\end{aligned}
$$

Hence, there is a one-one correspondence between representations and unordered pairs (u, v) of complementary factors of $m(m+1)$. (Since $m(m+1)$ is not square, u and v are always distinct.) Since m and $m+1$ are coprime, the number of factors of $m(m+1)$ is equal to $\tau(m)\tau(m+1)$, and so the number of representations is equal to $\frac{1}{2}\tau(m)\tau(m+1)$.

© Springer International Publishing Switzerland 2016
E.J. Barbeau, *University of Toronto Mathematics Competition (2001–2015)*, Problem Books in Mathematics,
DOI 10.1007/978-3-319-28106-3_11

Now consider the case that $m = 2001$. Since $2001 = 3 \times 23 \times 29$, $\tau(2001)=8$; since $2002 = 2 \times 7 \times 11 \times 13$, $\tau(2002) = 16$. Hence, the desired number of representations is 64.

2003:4. Show that n divides the integer nearest to

$$\frac{(n+1)!}{e}.$$

The solution to this problem appears in Chap. 4.

2004:3. Suppose that u and v are positive integer divisors of the positive integer n and that $uv < n$. Is it necessarily so that the greatest common divisor of n/u and n/v exceeds 1?

Solution 1. Let $n = ur = vs$. Then $uv < n \Rightarrow v < r, u < s$, so that $n^2 = uvrs \Rightarrow rs > n$. Let the greatest common divisor of r and s be g and the least common multiple of r and s be m. Then $m \le n < rs = gm$, so that $g > 1$.

Solution 2. Let $g = \gcd(u,v)$, $u = gs$ and $v = gt$. Then $gst \le g^2 st < n$ so that $st < n/g$. Now s and t are a coprime pair of integers, each of which divides n/g. Therefore, $n/g = dst$ for some $d > 1$. Therefore $n/u = n/(gs) = dt$ and $n/v = n/(gt) = ds$, so that n/u and n/v are divisible by d, and so their greatest common divisor exceeds 1.

Solution 3. $uv < n \implies nuv < n^2 \implies n < (n/u)(n/v)$. Suppose, if possible, that n/u and n/v have greatest common divisor 1. Then the least common multiple of n/u and n/v must equal $(n/u)(n/v)$. But n is a common multiple of n/u and n/v, so that $(n/u)(n/v) \le n$, a contradiction. Hence the greatest common divisor of n/u and n/v exceeds 1.

Solution 4. [D. Shirokoff] If n/u and n/v be coprime, then there are integers x and y for which $(n/u)x + (n/v)y = 1$, whence $n(xv + yu) = uv$. Since n and uv are positive, then so is the integer $xv + yu$. But $uv < n \implies 0 < xv + yu < 1$, an impossibility. Hence the greatest common divisor of n/u and n/v exceeds 1.

2004:9. Let $ABCD$ be a convex quadrilateral for which all sides and diagonals have rational length and AC and BD intersect at P. Prove that AP, BP, CP, DP all have rational length.

The solution to this problem appears in Chap. 8.

2005:7. Let $f(x)$ be a nonconstant polynomial that takes only integer values when x is an integer, and let P be the set of all primes that divide $f(m)$ for at least one integer m. Prove that P is an infinite set.

Solution 1. Let r be the greatest integer that divides all of the values $f(m)$, where $m \in \mathbb{Z}$, and let $g(x) = f(x)/m$. Then $g(x)$ is a nonconstant polynomial that satisfies the condition of the problem, and the set of all primes dividing at least one $f(m)$ is the union of the set of primes dividing at least one of $g(m)$ and the finites et of primes dividing r. Thus, it is enough to deal with $g(x)$.

Let $\{p_1, p_2, \ldots, p_k\}$ be a finite set of distinct primes. Since the greatest common divisor of the numbers $g(m)$ for $m \in \mathbb{Z}$ is 1, there is a set S_1 of numbers m for which $g(m)$ is not a multiple of p_1. Suppose, as an induction hypotheses, $1 \leq i \leq k-1$ and we have established that there is a set S_i of integers m for which $g(m)$ is divisible by none of $p_1, p_2, \ldots, p_i$. For any integers n and s, $f(n) \equiv f(n + sp_1p_2\cdots p_i) \bmod p_1p_2\cdots p_i$. Let $a \in S_i$; then $a + sp_1p_2\cdots p_i$ also belongs to S_i for all s.

There is an integer b for which $g(b)$ is not a multiple of p_{i+1}. Then neither is $g(n)$ a multiple of p_{i+1} when $n \equiv b \pmod{p_{i+1}}$. Since $p_1p_2\cdots p_i$ and p_{i+1} are coprime, $a + sp_1p_2\cdots p_i$ runs through a complete set of residues as s varies, and in particular is congruent to b for some values of s. Thus, S_i contains some number n for which $g(n)$ is not a multiple of p_{i+1}. Hence the set $S_{i+1} = \{m : g(m) \not\equiv 0 \forall p_j (1 \leq j \leq i+1)\}$ is nonvoid.

Therefore there exists an integer m for which $g(m)$ is divisible by none of $p_1, p_2, \ldots, p_k$, for an arbitrary set of primes, the result follows.

Solution 2. Let $f(x) = \sum_k^n a_k x_n$. The number $a_0 = f(0)$ is rational. Indeed, each of the numbers $f(0), f(1), \cdots, f(n)$ is an integer; writing these conditions out yields a system of $n+1$ linear equations with integer coefficients for the coefficients $a_0, a_1, \ldots, a_n$ whose determinant is nonzero. The solution of this equation consists of rational values. Hence all the coefficients of $f(x)$ are rational. Multiply $f(x)$ by the least common multiple of its denominators to get a polynomial $g(x)$ which takes integer values whenever x is an integer. Suppose, if possible, that values of $f(x)$ for integral x are divisible only by primes p from a finite set Q. Then the same is true of $g(x)$ for primes from a finite set P consisting of the primes in Q along with the prime divisors of the least common multiple, For each prime $p \in P$, select a positive integer a_p such that p^{a_p} does not divide $g(0)$. Let $N = \prod\{p^{a_p} : p \in P\}$. Then, for each integer u, $g(Nu) \equiv g(0) \not\equiv 0 \pmod{N}$. However, for all u, $g(Nu) = \prod p^{b_p}$, where $0 \leq b_p \leq a_p$. Since there are only finitely many numbers of this type, some number must be assumed by g infinitely often, yielding a contradiction. (*Alternatively:* one could deduce that $g(Nu) \leq N$ for all u and get a contradiction of the fact that $|g(Nu)|$ tends to infinity with u.)

Comment. There is another approach to showing that the coefficients of $f(x)$ are rational. The proof is by induction on the degree. Suppose that $p_k(x)$ is a polynomial of degree k assuming integer values at $x = n, n+1, \ldots, n+k$. Then, there are integers c_i for which

$$p_k(x) = c_{k,0}\binom{x}{k} + c_{k,1}\binom{x}{k-1} + \cdots + c_{k,k}\binom{x}{0}.$$

To see this, first observe that $\binom{x}{k}, \binom{x}{k-1}, \ldots, \binom{x}{0}$ constitute a basis for the vector space of polynomials of degree not exceeding k. So there exist *real* $c_{k,i}$

as specified. We prove by induction on k that the $c_{k,i}$ must in fact be integers. The result is trivial when $k = 0$. Assume its truth for $k \geq 0$. Suppose that

$$p_{k+1}(x) = c_{k+1,0}\binom{x}{k+1} + \cdots + c_{k+1,k+1}$$

takes integer values at $x = n, n+1, \ldots, n+k+1$. Then

$$p_{k+1}(x+1) - p_{k+1}(x) = c_{k+1,0}\binom{x}{k} + \cdots + c_{k+1,k}$$

is a polynomial of degree k which takes integer values at $n, n+1, \ldots, n+k$, and so $c_{k+1,0}, \ldots, c_{k+1,k}$ are all integers. Hence,

$$c_{k+1,k+1} = p_{k+1}(n) - c_{k+1,0}\binom{n}{k+1} - \cdots c_{k+1,k}\binom{n}{1}$$

is also an integer.

Solution 3. [R. Barrington Leigh] Let n be the degree of f. It can be established that if p be a prime and k a positive integer, then $f(x) \equiv f(x+p^{nk}) \pmod{p^k}$. The result of the problem holds for $n = 0$. Assume that it holds for $n = m-1$ and that $f(x)$ has degree m. Let $g(x) = f(x) - f(x-1)$, so that the degree of $g(x)$ is $m - 1$. Then, modulo p^k,

$$f(x + p^{nk}) - f(x) = \sum_{i=1}^{p^{nk}} g(x+i)$$

$$= \sum_{i=1}^{p^{(n-1)k}} (g(x+i) + g(x+i+p^{(n-1)k}) + \cdots + g(x+i+(p^k-1)p^{(n-1)k}))$$

$$\equiv \sum_{i=1}^{p^{(n-1)k}} p^k g(x+i) \equiv 0.$$

(Note that this does not require the coefficients to be integers.)

Suppose, if possible, that the set P of primes p that divide at least one value of $f(x)$ for integer x is finite, and that, for each $p \in P$, the positive integer a is chosen so that p^a does not divide $f(0)$. Let $q = \prod\{p^a : p \in P\}$. Then p^a does not divide $f(0)$, nor any of the values $f(q^n)$ for positive integer n, as these are all congruent modulo p^a. Since any prime divisor of $f(q^n)$ belongs to P, it must be that $f(q^n)$ is a divisor of q. But this contradicts the fact that $|f(q^n)|$ becomes arbitrarily large with n.

2007:2. Determine distinct positive integers a, b, c, d, e such that the five numbers a, b^2, c^3, d^4, e^5 constitute an arithmetic progression. (The difference between adjacent pairs is the same.)

Solution 1. One example is obtained by taking the arithmetic progression $(1, 9, 17, 25, 33)$ and multiplying by $3^{24}5^{30}11^{24}17^{20}$ to obtain

$$(a, b, c, d, e) = (3^{24}5^{30}11^{24}17^{20}, 3^{13}5^{15}11^{12}17^{10}, 3^8 5^{10}11^8 17^7, 3^6 5^8 11^6 17^5, 3^5 5^6 11^5 17^4).$$

Solution 2. [G. Siu] Let

$$(a, b, c, d, e) = (33 \times 97^{24} \times 65^{20}, 7 \times 97^{12} \times 65^{10}, 65^7 \times 97^8, 3 \times 97^6 \times 65^5, 97^5 \times 65^4).$$

Then

$$(a, b^2, c^3, d^4, e^5) = (33k, 49k, 65k, 81k, 97k)$$

where $k = 65^{20} \times 97^{24}$.

Comment. Two solvers found the loophole that the arithmetic progression itself could be constant, and gave the example $(a, b, c, d, e) = (n^{60}, n^{30}, n^{20}, n^{15}, n^{12})$ for an integer $n \geq 2$.

2007:6. Let $h(n)$ denote the number of finite sequences $\{a_1, a_2, \ldots, a_k\}$ of positive integers exceeding 1 for which $k \geq 1$, $a_1 \geq a_2 \geq \cdots \geq a_k$ and $n = a_1 a_2 \cdots a_k$. (For example, if $n = 20$, there are four such sequences $\{20\}$, $\{10, 2\}$, $\{5, 4\}$ and $\{5, 2, 2\}$ and $h(20) = 4$.)

Prove that

$$\sum_{n=1}^{\infty} \frac{h(n)}{n^2} = 1.$$

The solution to this problem appears in Chap. 4.

2007:9. Which integers can be written in the form

$$\frac{(x + y + z)^2}{xyz}$$

where x, y, z are positive integers?

Solution. Let $F(x, y, z)$ be the expression in question. Wolog, suppose that $x \leq y \leq z$. Suppose that $F(x, y, z) = n$. Then

$$nxyz = (x + y + z)^2 = (x + y)^2 + 2z(x + y) + z^2$$

from which z divides $(x + y)^2$. Let $w = (x + y)^2/z$. Then

$$F(x, y, w) = \frac{(zx + zy + (x + y)^2)^2}{zxy(x + y)^2} = \frac{(z + x + y)^2}{zxy} = F(x, y, z).$$

If $x + y \leq z$, then $w \leq x + y$. So, if there is a representation of n, we can find one for which $z \leq x + y$. Then

$$n = \frac{x}{yz} + \frac{y}{xz} + \frac{z}{xy} + 2\left(\frac{1}{x} + \frac{1}{y} + \frac{1}{z}\right)$$

$$\leq \frac{1}{z} + \frac{1}{x} + \left(\frac{1}{x} + \frac{1}{y}\right) + \frac{2}{x} + \frac{2}{y} + \frac{2}{z}$$

$$\leq \frac{7}{x} + \frac{3}{z}.$$

If $(x, y, z) = (1, 1, 1)$, then $n = 9$. Otherwise $n < 9$. We have that $F(9, 9, 9) = 1$, $F(4, 4, 8) = 2$, $F(3, 3, 3) = 3$, $F(2, 2, 4) = 4$, $F(1, 4, 5) = 5$, $F(1, 2, 3) = 6$, $F(1, 1, 2) = 8$.

However, $F(x, y, z)$ cannot equal 7. Supposing that $2 \leq x \leq y \leq z \leq x + y \leq 2y$, we have that

$$\frac{(x + y + z)^2}{xyz} \leq \frac{(x^2 + y^2) + 2x(y + z) + z^2 + 2yz}{2yz}$$

$$\leq \frac{x^2}{2yz} + \frac{y}{2z} + \frac{x(y + z)}{yz} + \frac{z^2}{2yz} + 2$$

$$\leq \frac{1}{2} + \frac{1}{2} + \frac{y(2z)}{yz} + \frac{z}{2y} + 2$$

$$= 1 + 2 + 1 + 2 = 6.$$

Now let $x = 1$ and $y \leq z \leq 1 + y$. Since $F(1, 1, 1)$ and $F(1, 1, 2)$ differ from 7, $y \geq 2$. But then

$$F(1, y, y) = \frac{(2y + 1)^2}{y^2} = 4 + \frac{4}{y} + \frac{1}{y^2} < 7,$$

and

$$F(1, y, y + 1) = \frac{(2y + 2)^2}{y(y + 1)} = \frac{4(y + 1)}{y} = 4 + \frac{4}{y} < 7.$$

Hence $F(x, y, z) = 7$ is not possible. Therefore, only the integers 1, 2, 3, 4, 5, 6, 8 can be represented.

2008:8. Let $b \geq 2$ be an integer base of numeration and let $1 \leq r \leq b-1$. Determine the sum of all r-digit numbers of the form

$$\overline{a_{r-1}a_{r-2}\ldots a_2 a_1 a_0} \equiv a_{r-1}b^{r-1} + a_{r-2}b^{r-2} + \cdots + a_1 r + a_0$$

whose digits increase strictly from left to right: $1 \leq a_{r-1} < a_{r-2} < \cdots < a_1 < a_0 \leq b - 1$.

Solution. The answer is $\overline{12 \ldots r}\binom{b}{r+1}$.

Let n_s denote the s-digit number $\overline{12 \ldots s}$ for each positive integer $s \leq b - 1$. We use a double induction and show that for $b - 1 \geq k \geq r$, the sum of all the r-digit numbers whose units digit is k is $n_r \binom{k}{r}$.

When $r = 1$, there is only one number ending in k, namely k itself, and the desired sum is $k = 1 \times \binom{k}{1}$. For arbitrary r and $k = r$, the only eligible number is $\overline{12 \ldots r} = n_r$.

Suppose that the result holds up to $r - 1$ and up to k. The r-digit numbers that end in $k + 1$ fall into two classes: (a) those whose second last digit does not exceed $k - 1$ and (b) those whose last two digits are k and $k + 1$. There are $\binom{k-1}{r-1}$ numbers in class (a) and $\binom{k-1}{r-2}$ numbers in class (b).

Since the numbers in class (a) can be formed by taking all the numbers with r-digits and last digit k and adding 1 to each such number, the sum of the numbers in class (a) is equal to

$$n_r \binom{k}{r} + \binom{k-1}{r-1}.$$

The numbers in class (b) can be found by taking all the $(r-1)$-digit numbers ending in k, multiplying them by b and adding $k+1$ to each such number. The sum of these numbers is

$$bn_{r-1}\binom{k}{r-1} + \binom{k-1}{r-2}(k+1) = (n_r - r)\binom{k}{r-1} + \binom{k-1}{r-2}(k+1).$$

Therefore the sum of all the numbers in classes (a) and (b) is

$$n_r\binom{k}{r} + \binom{k-1}{r-1} + n_r\binom{k}{r-1} - r\binom{k}{r-1} + (k+1)\binom{k-1}{r-2}$$

$$= n_r\left[\binom{k}{r} + \binom{k}{r-1}\right] + \left[\binom{k-1}{r-1} + (k+1)\binom{k-1}{r-2} - r\binom{k}{r-1}\right]$$

$$= n_r\binom{k+1}{r} + 0 = n_r\binom{k+1}{r}.$$

Therefore the sum of all the r-digit numbers with increasing digits is

$$n_r \sum_{k=r}^{b-1} \binom{k}{r} = n_r\binom{b}{r+1}.$$

2009:3. For each positive integer n, let $p(n)$ be the product of all positive integral divisors of n. Is it possible to find two distinct positive integers m and n for which $p(m) = p(n)$?

Solution. The answer is **no.** The function $p(n)$ is one-one. Observe that, since each divisor d can be paired with n/d (which are distinct except when n is square and d is its square root), $p(n) = n^{\tau(n)/2}$, where $\tau(n)$ is the number of divisors of n.

Suppose that $p(m) = p(n)$ and that q is any prime. Let the highest powers of q dividing m and n be q^u and q^v respectively. The exponent of the highest power of q that divides $p(m)$ is $(u\tau(m))/2$ and the highest power of q that divides $p(n)$ has exponent $(v\tau(n))/2$. (The exponents are indeed integers, since $\tau(n)$ is odd if and only if all u are even.)

Since $p(m) = p(n)$, it follows that $u\tau(m) = v\tau(n)$. If $u < v$ for some prime q, it is so for all primes q, so that $m < n$ and $\tau(m) < \tau(n)$. Thus, $u < v$ is impossible, as is $u > v$. Therefore $u = v$ for each prime q. But then this means that $m = n$.

2009:6. Determine all solutions in nonnegative integers (x, y, z, w) to the equation

$$2^x 3^y - 5^z 7^w = 1.$$

Solution. By parity considerations, we must have that $x \geq 1$. Suppose, first of all, that $y \geq 1$, $z \geq 1$ and $w \geq 1$. Since $2^x 3^y = 6^y 2^{x-y} \equiv 2^{x-y} \equiv 1$ (mod 5), it follows that $x \equiv y$ (mod 4), so that x and y have the same parity.

Since $2^x 3^y \equiv 1 \pmod 7$, it follows that $(x, y) \equiv (0,0), (1,4), (2,2), (3,0), (4,4)$, $(5,2) \pmod 6$. Therefore, x and y are both even. Let $x = 2u$ and $y = 2v$. Then

$$5^z 7^w = 2^{2u} 3^{2v} - 1 = (2^u 3^v - 1)(2^u 3^v + 1).$$

The two factors on the right must be coprime (being consecutive odd numbers), so that one of them is a power of 5 and the other is a power of 7. Thus $2^u 3^v \pm 1 = 5^z$ and $2^u 3^v \mp 1 = 7^w$, whence

$$7^w - 5^z = \mp 2.$$

Observe that 7^w is congruent, modulo 25, to one of $1, 7, -1, -7$, so that $7^w \mp 2$ is never divisible by 25. Therefore, $z = 1$ and so $w = 1$ and we obtain the solution $(x, y, z, w) = (2, 2, 1, 1)$.

Now we consider the cases where at least one of y, z, w vanishes.

Suppose that $y = 0$. Then $2^x - 1 = 5^z 7^w$. If x is even, then the left side is divisible by 3. Therefore x is odd. But if x is odd, then the left side is not divisible by 5. Therefore $z = 0$ and $2^x - 7^w = 1$. This implies either that $w = 0$ or that x is divisible by 3. There are two immediate possibilities: $(x, w) = (1, 0)$ and $(x, w) = (3, 1)$. Suppose that $x = 3r$ where $r \geq 2$. Then

$$7^w = (2^r)^3 - 1 = (2^r - 1)(2^{2r} + 2^r + 1),$$

so that both factors on the right are nontrivial powers of 7. But it is not possible for both factors of the right side to be divisible by 7, and we obtain no further solutions.

Therefore, when $y = 0$, we have only the solutions $(x, y, z, w) = (1, 0, 0, 0)$, $(3, 0, 0, 1)$.

It is not possible for x and y to both exceed 0 while $z = 0$, since otherwise $2^x 3^y - 5^z 7^w$ would be congruent to -1 (modulo 3).

Finally, let x and y both exceed 0 and let $w = 0$. Since $5^z + 1 \equiv 2$ (mod 4), we must have that $x = 1$. Observe that $5^z + 1 \equiv 0 \pmod 9$ only when $z \equiv 3 \pmod 6$. However, when $z \equiv 3 \pmod 6$, $5^z + 1 \equiv 0 \pmod 7$. Therefore, either $5^z + 1$ is not divisible by 9, or it is divisible by 7. In either case, it cannot be of the form $2^x 3^y$ when $y \geq 2$. Therefore, $y = 1$. This leads to the sole solution $(x, y, z, w) = (1, 1, 1, 0)$.

In all, there are four integral solutions to the equation, namely

$$(x, y, z, w) = (2, 2, 1, 1), (1, 1, 1, 0), (3, 0, 0, 1), (1, 0, 0, 0).$$

2009:9. Let p be a prime congruent to 1 modulo 4. For each real number x, let $\{x\} = x - \lfloor x \rfloor$ denote the fractional part of x. Determine

$$\sum \left\{ \left\{ \frac{k^2}{p} \right\} : 1 \leq k \leq \frac{1}{2}(p - 1) \right\}.$$

Solution. When x is not divisible by p, then $x \not\equiv -x \pmod{p}$, since p is odd. Also, $x^2 \equiv y^2 \pmod{p}$ if and only if $x \equiv y$ or $x \equiv -y \pmod{p}$, so that the mapping $x \to x^2$ is a two-one mapping on $\mathbb{Z}_p^*$ (the integers modulo p that are coprime with p); thus, precisely $\frac{1}{2}(p-1)$ elements of $\mathbb{Z}_p^*$ are squares. We first show that m is a square modulo p if and only if $p - m$ is a square modulo p.

If s is a square in $\mathbb{Z}_p^*$ with $r^2 = s$, then $s^{-1} = (r^{-1})^2$ is also a square, so we can split the squares into disjoint sets of inverses. $s = s^{-1}$ can occur only if $s^2 = 1$ or $s = \pm 1$ in $\mathbb{Z}_p$. Thus, these disjoint sets contain precisely two elements when the square is not 1 nor -1, if applicable. Since there are an even number $\frac{1}{2}(p-1)$ of squares and 1 is a square, there must be another singleton set of inverse squares, and this can consist only of -1. Hence -1 is a square in $\mathbb{Z}_p^*$. Therefore, if m is a square in $\mathbb{Z}_p^*$, so also is $-m = (-1)m$. In other words, m is a square modulo p if and only if $p - m$ is a square modulo p.

The set $\{k^2 : 1 \le k \le \frac{1}{2}(p-1)\}$ contains each nonzero square exactly once, and so there are $\frac{1}{4}(p-1)$ pairs of squares of the form $\{m, p-m\}$ where $1 \le m \le \frac{1}{2}(p-1)$. If $k^2 \equiv r$ with $1 \le r \le p-1$, then $\{k^2/p\} = r/p$. Therefore

$$\sum_{k=1}^{\frac{1}{2}(p-1)} \left\{\frac{k^2}{p}\right\} = \sum \left\{\left(\frac{m}{p} + \frac{p-m}{p}\right) : 1 \le m \le \frac{1}{2}(p-1), m \text{ a square } (\bmod\ p)\right\}$$

$$= \frac{p-1}{4}.$$

2011:10. Suppose that p is an odd prime. Determine the number of subsets S contained in $\{1, 2, \ldots, 2p-1, 2p\}$ for which (a) S has exactly p elements, and (b) the sum of the elements of S is a multiple of p.

The solution to this problem appears in Chap. 2.

2012:2. Suppose that f is a function defined on the set $\mathbb{Z}$ of integers that takes integer values and satisfies the condition that $f(b) - f(a)$ is a multiple of $b - a$ for every pair a, b, of integers. Suppose also that p is a polynomial with integer coefficients such that $p(n) = f(n)$ for infinitely many integers n. Prove that $p(x) = f(x)$ for every positive integer x.

Solution. Let $q(x) = f(x) - p(x)$ for each integer x. The function $q(x)$ vanishes on an infinite subset M of integers and $q(a) - q(b)$ is divisible by $a - b$ for every pair a, b of integers, since the same is true of $p(x)$. (If $p(x) = \sum c_i x^i$, then $p(a) - p(b) = \sum c_i(a^i - b^i) = (a - b)\sum c_i(a^{i-1} + a^{i-2}b + \cdots + ab^{i-2} + b^{i-1})$.)

Let n be any integer. Then $q(n) = q(n) - q(m)$ is divisible by $n - m$ for each $m \in M$. But then either $q(n) = 0$ or $|q(n)| \ge |n - m|$ for infinitely many integers m. However, the latter case cannot occur.

Comment. The function $f(x)$ itself need not be a polynomial. For example, we might have $f(x) = x + \sin \pi x$.

2012:4. (a) Let n and k be positive integers. Prove that the least common multiple of $\{n, n+1, n+2, \ldots, n+k\}$ is equal to

$$rn\binom{n+k}{k}$$

for some positive integer r.

(b) For each positive integer k, prove that there exist infinitely many positive integers n, for which the number r defined in part (a) is equal to 1.

Solution 1. (a) Observe that

$$n\binom{n+k}{k} = \frac{n(n+1)(n+2)\cdots(n+k)}{(1)(2)(3)\cdots(k)}.$$

Let p be an arbitrary prime and suppose that p^b is the highest power of p that divides any of n, $n+1$, $\ldots$, $n+k$; thus, p^b is the power of p that occurs in the prime factorization of the least common multiple of n, $n+1$, $\ldots$, $n+k$. We need to show that p divides $n\binom{n+k}{k}$ to no greater a power than p^b.

Suppose that $n+u$ is divisible by p^b where $0 \le u \le k$. Write $n\binom{n+k}{k} = (n+u) \times A \times B$ where

$$A = \frac{(n+u+1)(n+u+2)\cdots(n+k)}{(1)(2)(3)\cdots(k-u)}$$

and

$$B = \frac{(n+u-1)(n+u-2)\cdots(n)}{(k-u+1)(k-u+2)\cdots(k)}.$$

Since $n+u+i \equiv i \pmod{p^b}$ it follows that p divides $n+u+i$ and i to the same power, for $1 \le i \le k-u$. Therefore A, written in lowest terms, has a numerator that is not divisible by p. Since $n+u-i \equiv -i \pmod{p^b}$ for $1 \le i \le u$ and $k-u+1$, $k-u+2$, $\ldots$, k constitutes a set of u consecutive integers, for each power p^a with $a \le b$, p^a divides a number among $k-u+1$, $\ldots$, k at most once more (either $\lfloor u/p^a \rfloor$ or $\lfloor u/p^a \rfloor + 1$) than it divides one of the numbers -1, -2, $\ldots$, $-u$ or one of the numbers $n+u-1$, $n+u-2$, $\ldots$, n. It follows that the power of p that divides the denominator of B cannot exceed the power that divides the numerator of B by more than b. Since $n+u$ is exactly divisible by p^b, it follows that the highest power of p that divides $n\binom{n+k}{k}$ is no greater than b. The desired result follows.

(b) Let $n = r \times k!$ for any positive integer r. Then $n\binom{n+k}{k} = rn(n+1)(n+2)\cdots(n+k)$ which is clearly a common multiple of n, $n+1$, $\ldots$, $n+k$. Since this number is a divisor of the least common multiple of n, $n+1$, $\ldots$, $n+k$, it must be equal to the least common multiple.

Solution 2. [J. Zung] (a) For a prime p, let $f_p(m)$ denote the exponent of the highest power of p that is a divisor of the integer m. It is required to prove that

$$f_p(n(n+1)\cdots(n+k)) - f_p(k!) = f_p\left(n\binom{n+k}{k}\right)$$

$$\leq f_p(\text{lcm}\,(n, n+1, \ldots, n+k))$$

$$= \max\{f_p(m) : n \leq m \leq n+k\}.$$

Suppose that M is the maximum of $f_p(m)$ for $n \leq m \leq n+k$. Then $f_p(n(n+1)\cdots(n+k))$ is the sum over i from 1 to M of the number of multiples of p^i that occur in the set $\{n, n+1, \ldots, n+k\}$. This is at most once more than the number of multiples of p^i that occur in the set $\{1, 2, \ldots, k\}$ with equality occurring when n is a multiple of p^i. Therefore

$$f_p(n(n+1)\cdots(n+k)) \leq \sum_{1 \leq i \leq M}\left(\left\lfloor\frac{k}{p^i}\right\rfloor + 1\right) = f_p(k!) + M,$$

which yields the desired result.

(b) Equality occurs when n is a multiple of $p^{f_p(k!)}$ for every prime p.

2013:1. Let a be an odd positive integer exceeding 3, and let n be a positive integer. Prove that

$$a^{2^n} - 1$$

has at least $n+1$ distinct prime divisors.

(b) When $a = 3$, determine all the positive integers n for which the assertion in (a) is false.

Solution 1. (a) For $n = 1$, note that $a^2 - 1 = uv$, where $u = a - 1$ and $v = a + 1$. Since u and v are consecutive even integers, one of them is the product of 2 and an odd integer. Thus, $a^2 - 1$ has at least two prime divisors. We complete the proof by induction. Suppose it holds for the exponent n. Observe that

$$a^{2^n} - 1 = (a^{2^{n-1}} + 1)(a^{2^{n-1}} - 1).$$

The second factor on the right, by the induction hypothesis, has at least n distinct prime divisors. The first factor, being congruent to 2 modulo 4, must have an odd divisor exceeding 1. Any odd prime that divides the first factor cannot divide the second, and so there must be at least $n+1$ prime factors in all.

(b) Observe that $3^2 - 1 = 2^3$, $3^{2^2} - 1 = 3^4 - 1 = 2^4 \times 5$,

$$3^{2^3} - 1 = (3^4 - 1)(3^4 + 1) = (2^4 \times 5) \times (2 \times 41) = 2^5 \times 5 \times 41$$

and

$$3^{2^4} - 1 = (3^8 - 1)(3^8 + 1) = (2^5 \times 5 \times 41) \times (2 \times 17 \times 193) = 2^6 \times 5 \times 17 \times 41 \times 193.$$

By induction, as in (a), it can be shown that $3^{2^n} - 1$ has $n+1$ distinct prime factors when $n \geq 4$. It has exactly n distinct prime factors when $n = 1, 2, 3$.

Solution 2. [J. Love] (a) Observe that, for each positive integer n,

$$a^{2^n} - 1 = (a-1)(a+1)(a^2+1)\ldots(a^{2^{n-1}}+1).$$

This quantity is divisible by 2. Since a is odd, $a^2 \equiv 1 \bmod 4$, so that the last $n-1$ terms of the product on the right is equal to twice an odd integer, and therefore has at least one odd prime divisor. Since $a > 3$, at least one of $a-1$ and $a+1$ is equal to twice an odd prime.

It remains to show that no two of the factors on the right side can be divisible by the same odd prime. If p is an odd prime divisor of $a-1$, then each of the remaining factors is congruent to 2 modulo p, If $i < j$ and p divides $a^{2^i}+1$ and $a^{2^j}+1$, then, since $a^{2^i}+1$ divides $a^{2^j}-1$, then so does p. But then p divides $2 = (a^{2^j}+1)-(a^{2^j}-1)$, which is false. Therefore, $a^{2^n}-1$ is divisible by 2 and n odd primes, distinct divisors of a^2-1 and $a^{2^i}+1$ for $1 \le i \le n-1$.

(b) As in (a), we can prove that 2 divides $N \equiv 32^n - 1$ and that distinct odd primes divide each of the $n-1$ factors 3^2+1, 3^4+1, ..., $3^{2^{n-1}}+1$ of N. Thus, there are at least n distinct primes dividing N. However, $3^{2^3}+1 = 6562 = 2 \times 17 \times 193$ is divisible by 2 odd primes, so that, when $n \ge 4$, N is divisible by $n+1$ distinct primes. However, since $3^2-1 = 2^3$, $3^4-1 = 2^4 \times 5$ and $3^8-1 = 6560 = 2^5 \times 5 \times 41$, the assertion in (a) fails for $a = 3$ and $n = 1, 2, 3$.

2013:4. Let S be the set of integers of the form $x^2 + xy + y^2$, where x and y are integers.

(a) Prove that any prime p in S is either equal to 3 or is congruent to 1 modulo 6.

(b) Prove that S includes all squares.

(c) Prove that S is closed under multiplication.

Solution. (a) Let $f(x,y) = x^2+xy+y^2$. Then $3 = f(1,1)$ is representable. It is straightforward to establish that 2 is not representable. (If $x^2+xy+y^2 = 2$, then $(2x+y)^2+3y^2 = 8$.) Let p be a prime exceeding 3 for which $p = f(u,v)$ for some integers u and v. Then, modulo 3,

$$0 \not\equiv p \equiv 4p = 4(u^2 + uv + v^2) = (2u+v)^2 + 3v^2 \equiv (2u+v)^2 \equiv 1.$$

Since p is odd, the result follows. (Alternatively, modulo 3, we have that $0 \not\equiv p = (u-v)^2 + 3uv \equiv (u-v)^2 \equiv 1$.)

(b) Observe that $a^2 = f(a,0) = f(a,-a)$ for each integer a.

(c) Observe that $x^2 + xy + y^2 = (x - y\omega)(x - y\omega^2)$, where ω is an imaginary cube root of unity, i.e., $1 + \omega + \omega^2 = 0$ (this can be deduced from the factorization of $x^3 - y^3$ into linear factors over $\mathbb{C}$). Then, since $\omega^2 = -1 - \omega$ and $\omega^3 = 1$.

$$(x^2 + xy + y^2)(u^2 + uv + v^2)$$
$$= [(x - y\omega)(x - y\omega^2)][(u - v\omega)(u - v\omega^2)]$$
$$= [(x - y\omega)(u - v\omega)][(x - y\omega^2)(u - v\omega^2)]$$
$$= [xu - (xv + yu)\omega + yv\omega^2][xu - (xv + yu)\omega^2 + yv\omega]$$
$$= [(xu - yv) - (xv + yu + yv)\omega(xu - yv) - (xv + yu + yv)\omega^2].$$

It is readily checked that

$$f(x,y)f(u,v) = f(xu - yv, xv + yu + yv).$$

2014:2. For a positive integer N written in base 10 numeration, N' denotes the integer with the digits of N written in reverse order. There are pairs of integers (A, B) for which A, A', B, B' are all distinct and $A \times B = B' \times A'$. For example,

$$3516 \times 8274 = 4728 \times 6153.$$

   (a) Determine a pair (A, B) as described above for which both A and B have two digits, and all four digits involved are distinct.
   (b) Are there any pairs (A, B) as described above for which A has two and B has three digits?

Solution. (a) Let $A = 10a + b$ and $B = 10u + v$ where $1 \le a, b, u, v \le 9$. The condition $AB = B'A'$ leads to $99(au - bv) = 0$, so that a necessary condition for the property to hold is $au = bv$. There are many possibilities, including $(a, b; u, v) = (1, 3; 6, 2)$. Thus, for example, $13 \times 62 = 26 \times 31$. (The possibilities include $(A, B) = (12, 42), (12, 63), (13, 62), (12, 84), (14, 82), (23, 64), (24, 63), (24, 84), (23, 96), (26, 93), (34, 86), (46, 96), (48, 63)$.)

   (b) Let $A = 10a + b$ and $B = 100u + 10v + w$, where $abuw \ne 0$ and $0 \le a, b, u, v, w \le 9$. The condition $AB = B'A'$ leads to

$$111(au - bw) + 10[a(v - w) + b(u - v)] = 0.$$

Therefore $a(v - w) + b(u - v)$ must be a multiple of 111. If the two terms of this sum differ in sign, then the absolute value of the sum cannot exceed $9 \times 9 = 81$. If the two terms of the sum have the same sign, then the signs of $v - w$ and $u - v$ must be the same, and the absolute value of the sum cannot exceed

$$|a(v - w) + b(u - v)| \le 9|(v - w) + (u - v)| = 9|u - w| = 81.$$

Therefore $a(v - w) + b(u - v) = 0$, so that $au = bw$. If $v - w$ and $u - v$ do not vanish, they must have opposite signs. These conditions are necessary and sufficient for an example. Trial and error leads to $(a, b; u, v, w) = (1, 2; 2, 3, 1)$. Thus, we obtain

$$12 \times 231 = 2772 = 132 \times 21.$$

In fact, we can replace these numbers by multiples that involve no carries to obtain the examples $(12s, 231t)$ where $1 \le s \le 4$ and $1 \le t \le 3$. Other examples of pairs are $(13, 682)$, $(28, 451)$.

Comment. At a higher level, we have that

$$992 \times 483 \times 156 = 651 \times 384 \times 299 = 2^7 \times 3^2 \times 7 \times 13 \times 23 \times 31.$$

2014:3. Let n be a positive integer. A finite sequence $\{a_1, a_2, \ldots, a_n\}$ of positive integers a_i is said to be *tight* if and only if $1 \le a_1 < a_2 < \cdots < a_n$, all $\binom{n}{2}$ differences $a_j - a_i$ with $i < j$ are distinct, and a_n is as small as possible.

(a) Determine a tight sequence for $n = 5$.

(b) Prove that there is a polynomial $p(n)$ of degree not exceeding 3 such that $a_n \le p(n)$ for every tight sequence $\{a_i\}$ with n entries.

Solution. (a) A sequence having all differences different remains so if the same integer is added or subtracted from each term. Therefore $a_1 = 1$ for any tight sequence. Since there are $\binom{5}{2} = 10$ differences for a tight sequence with five entries, $a_5 \ge 11$. We show that $a_5 = 11$ cannot occur.

Suppose, if possible, that $a_1 = 1$ and $a_5 = 11$. Since each of the differences from 1 to 10, inclusive, must occur, and since the largest difference 10 is the sum of the four differences of adjacent entries, the differences of adjacent entries must be 1, 2, 3, 4 in some order. The difference 1 cannot be next to either of the differences 2 or 3, for then there a difference between nonadjacent entries would be 3 or 4, respectively. Thus 1 is the difference of either the first two or the last two entries and the adjacent difference would be 4. But then $a_3 - a_1 = a_5 - a_3 = 5$, which is not allowed. Since there is no other possibility, $a_5 \ge 12$.

Trying $a_5 = 12$, we find that the sequence $\{1, 2, 5, 10, 12\}$ is tight.

(b) The strategy is to construct a sequence that satisfies the difference condition such that the differences between pairs of terms have different congruence classes with respect to some modulus, dependent only on the number of terms between them. Thus, we make the difference between adjacent terms congruent to 1, between terms separated by one entry congruent to 2, etc. To this end, let $a_1 = 1$ and

$$a_{i+1} = a_i + 1 + (i - 1)(n - 1)$$

for $1 \le i \le n - 1$. Then,

$$a_i = i + \frac{1}{2}(i - 1)(i - 2)(n - 1).$$

In particular,

$$a_n = n + \frac{1}{2}(n - 1)^2(n - 2) = \frac{1}{2}(n^3 - 4n^2 + 5n - 2).$$

When $1 \le i < j \le n$, we have that

$$a_j - a_i = (a_j - a_{j-1}) + (a_{j-1} - a_{j-2}) + \cdots + (a_{i+1} - a_i) \equiv j - i$$

modulo $(n-1)$. Furthermore,

$$(a_{j+1} - a_{i+1}) - (a_j - a_i) = (a_{j+1} - a_j) - (a_{i+1} - a_i) = (j-i)(n-1) > 0.$$

It follows that all the differences are distinct.

Any tight sequence with n terms must have its largest term no greater than $p(n) = \frac{1}{2}(n^3 - 4n^2 + 5n - 2)$.

Comment. This upper bound is undoubtedly too generous and it may be that there is an upper bound quadratic in n.

2014:5. Let n be a positive integer. Prove that

$$\sum_{k=1}^{n} \frac{1}{k\binom{n}{k}} = \sum_{k=1}^{n} \frac{1}{k2^{n-k}} = \frac{1}{2^{n-1}} \sum_{k=1}^{n} \frac{2^{k-1}}{k} = \frac{1}{2^{n-1}} \sum \left\{ \frac{\binom{n}{k}}{k} : k \text{ odd}, 1 \le k \le n \right\}.$$

The solution to this problem appears in Chap. 4.

2015:6. Using the digits 1, 2, 3, 4, 5, 6, 7, 8, each exactly once, create two numbers and form their product. For example, $472 \times 83{,}156 = 39{,}249{,}632$. What are the smallest and the largest values such a product can have?

Solution. Observe that, when $a > b > 0$, $c > 0$, then $(10b + c)a - (10a + c)b = c(a - b) > 0$. If a has as many digits as $10b + c$ (where $c \le 9$) and $a > 10b + c > b$, then taking the last digits from $10b + c$ and appending it to a gives the product $(10a+c)b < (10b+c)a$. From this, we see that if the two numbers have unequal numbers of digits or the same number of digits, removing the last digit from the smaller and appending it to the larger, will result in a smaller product of the pair. Therefore, the smallest product will occur when one number has a single digit and the other seven digits. For both factors, the digits will appear in increasing order. It is straightforward to see that the smallest product is $1 \times 2{,}345{,}678 = 2{,}345{,}678$.

Similarly, if the two factors have unequal numbers of digits, removing the last digit from the larger factor and appending it to the smaller will result in a larger product for the pair. Therefore, the largest product will occur when both factors have four digits and the digits appear in decreasing order. Observe the $(a + c)(b + d) - (a + d)(b + c) = (a - b)(d - c)$. If $a > b$, this will be positive if and only if $c < d$. One number begins with 8 and the other with either 7 or 6. But since $\overline{87pq} \times \overline{6mcd} < \overline{8mrs} \times \overline{67pq} < \overline{8mrs} \times \overline{76pq}$, one number begins with 8 and the other with 7. Applying the result further, we deduce that each digit after the first in the number beginning with 8 is less than the corresponding digit after the first in the number beginning with 7. Therefore, we deduce that the largest product is $8531 \times 7642 = 65{,}193{,}902$.

Definitions, Conventions, Notation and Basics

1. General

1.1. Wolog = Without loss of generality

1.2. Kronecker delta. For indices (i, j), the Kronecker delta function is defined by

$$\delta_{ij} = \begin{cases} 1 & \text{if } i = j; \\ 0 & \text{if } i \neq j. \end{cases}$$

1.3. Logarithm. The symbol log refers to the natural logarithm, or logarithm to base e. The logarithm to base b will be indicated as $\log_b$.

1.4. Floor and ceiling. For any real number x, the *floor* of x, denoted by $\lfloor x \rfloor$, is the largest integer that does not exceed x. The *ceiling* of x, denoted by $\lceil x \rceil$, is the smallest integer that is not less than x.

2. Algebra

2.1. Binomial symbol. Let x be any real number and n be a positive integer. The *binomial symbol* is defined as

$$\binom{x}{n} = \frac{x(x-1)\cdots(x-n+1)}{n!}.$$

A fundamental identity is:

$$\binom{x+1}{n} = \binom{x}{n} + \binom{x}{n-1}$$

for $n \geq 1$.

© Springer International Publishing Switzerland 2016
E.J. Barbeau, *University of Toronto Mathematics Competition (2001–2015)*, Problem Books in Mathematics,
DOI 10.1007/978-3-319-28106-3

2.2. Roots of unity. The solutions of the equation $z^n - 1 = 0$, known as the nth roots of unity, are the complex numbers $\cos(2k\pi/n) + i\sin(2k\pi/n) = e^{(2k\pi/n)i}$ where $0 \le k \le n - 1$. This is the set $\{1, \zeta, \zeta^2, \ldots, \zeta^{n-1}\}$ where $\zeta = \cos(2\pi/n) + i\sin(2\pi/n)$ is a fundamental nth root of unity. Every nth root of 1 except 1 itself satisfies the equation $1 + z + z^2 + \cdots + z^{n-1} = 0$.

2.3. Fundamental theorem of algebra. Every polynomial $p(z)$ with complex coefficients of positive degree n can be written as a product

$$a(z - r_1)^{m_1}(z - r_2)^{m_2} \cdots (z - r_k)^{m_k},$$

where $m_1 + m_2 + \cdots + m_k = n$. For each i, the number r_i is a root of $p(z)$ with multiplicity m_i.

2.4. Symmetric function of the roots. Let $r_1, r_2, \ldots, r_n$ be the n roots of the polynomial $z^n + a_{n-1}z^{n-1} + a_{n-2}z^{n-2} + \cdots + a_1 z + a_0$. Then $r_1 + r_2 + \cdots + r_n = -a_{n-1}$, $r_1 r_2 \cdots r_n = (-1)^n a_0$, and, for $1 \le k \le n$, the sum of all $\binom{n}{k}$ k-fold products of the roots is $(-1)^k a_{n-k}$.

2.5. Sum of powers of the roots. With the same notation as before, let $p_k = r_1^k + r_2^k + \cdots + r_n^k$ be the sum of the kth powers of the roots for nonnegative integers k. Thus $p_0 = n$, $p_1 = -a_{n-1}$, $p_2 = a_{n-1}^2 - 2a_{n-2}$. More generally, when $1 \le k \le n - 1$,

$$p_k + a_{n-1}p_{k-1} + a_{n-2}p_{k-2} + \cdots + a_{n-k+1}p_1 + ka_{n-k} = 0.$$

For $k = n + r \ge n$, we have that

$$p_{n+r} + a_{n-1}p_{n+r-1} + a_{n-2}p_{n+r-2} + \cdots + a_1 p_{r+1} + a_0 p_r = 0.$$

2.6. Rational function. A *rational function* is a ratio of two polynomials. The roots of the numerator are the *roots* of the rational function, and the roots of the denominator are the *poles* of the rational function.

2.7. Greatest common divisor and Euclidean algorithm. A *greatest common divisor* of two polynomials is a polynomial of maximum degree which divides both the polynomials. The polynomials are *coprime* iff their greatest common divisors are constants. Given any two polynomials $f(z)$ and $g(z)$, we can determine polynomials $q(z)$ and $r(z)$, with the degree of r less than that of q for which $f(z) = q(z)g(z) + r(z)$. A greatest common divisor of the pair $(f(z), g(z))$ is equal to a greatest common divisor of $(g(z), r(z))$. By dividing $g(z)$ by $r(z)$, and continuing the process, we get a succession of equations

$$f(z) = q(z)g(z) + r(z)$$
$$g(z) = q_1(z)r(z) + r_1(z)$$
$$r(z) = q_2(z)r_1(z) + r_2(z)$$
$$\cdots$$
$$r_{i-1}(z) = q_{i+1}(z)r_i(z) + r_{i+1}(z)$$

$\cdots$

$$r_{k-1}(z) = q_{k+1}(z)r_k(z),$$

where the degrees of the remainders strictly decrease and the process eventually terminates with an exact division. The polynomial $r_k(z)$ is a greatest common divisor of $f(z)$ and $g(z)$. This method of obtaining a greatest common divisor is the *Euclidean algorithm*.

2.8. Trinomial theorem. The *trinomial theorem* states that

$$(a + b + c)^n = \sum \left\{ \frac{(i+j+k)!}{i!j!k!} a^i b^j c^k : i, j, k \geq 0, i + j + k = n \right\}.$$

3. Inequalities

3.1. Arithmetic-geometric means inequality. Let $a_1, a_2, \cdots, a_n$ be nonnegative real numbers. Then

$$(a_1 a_2 \cdots a_n)^{1/n} \leq \frac{a_1 + a_2 + \cdots + a_n}{n},$$

with equality if an only if $a_1 = a_2 = \cdots = a_n$. This can be generalized to a weighted version. Suppose that $\lambda_1, \lambda_2, \ldots, \lambda_n$ are positive real numbers for which $\lambda_1 + \lambda_2 + \cdots + \lambda_n = 1$. Then

$$a_1^{\lambda_1} a_2^{\lambda_2} \cdots a_n^{\lambda_n} = \lambda_1 a_1 + \lambda_2 a_2 + \cdots + \lambda_n a_n.$$

3.2. Power means inequalities. Let $a_1, a_2, \ldots, a_n$ be nonnegative real numbers, and let m be a nonzero integer. The mth power mean M_m is defined as

$$M_m = (a_1^m + a_2^m + \cdots + a_n^m)^{1/m}.$$

We also define M_0 to be the geometric mean $(\prod(a_i))^{1/n}$ of the a_i. Then for $r < s$, we have that

$$\min(a_i) \leq M_r \leq M_s \leq \max a_i.$$

M_2 is the *root-mean-square*; M_1 is the *arithmetic mean*; M_0 is the *geometric mean*; and M_{-1} is the *harmonic mean* of the a_i.

3.3. Cauchy-Schwarz Inequality. Let a_i and b_i be real numbers, for $1 \leq i \leq n$. Then

$$(a_1 b_1 + a_2 b_2 + \cdots + a_n b_n)^2 \leq (a_1^2 + a_2^2 + \cdots + a_n^2)(b_1^2 + b_2^2 + \cdots + b_n^2),$$

with equality if and only if $a_i = b_i$ for each i. There is also an integral version:

$$\left(\int_a^b f(x)g(x)dx \right)^2 \leq \left(\int_a^b f(x)^2 dx \right) \left(\int_a^b g(x)^2 dx \right).$$

3.4. Jensen Inequality. There are two versions, one for convex and one for concave functions.

A real-valued function defined on $\mathbb{R}$ is *convex* if and only if, for any real x and y and for $0 \leq t \leq 1$, then $f((1-t)x + ty) \leq (1-t)f(x) + tf(y)$. Then for any set $\{x_1, x_2, \ldots, x_n\}$ of real numbers and any set of positive reals $\{\lambda_1, \lambda_2, \ldots, \lambda_n\}$ with $\lambda_1 + \lambda_2 + \cdots + \lambda_n = 1$, we have that

$$f(\lambda_1 x_1 + \lambda_2 x_2 + \cdots + \lambda_n x_n) \leq \lambda_1 f(x_1) + \lambda_2 f(x_2) + \cdots + \lambda_n f(x_n).$$

A twice-differential function on an open interval whose second derivative is nonnegative is convex.

A real-valued function defined on $\mathbb{R}$ is *concave* if and only if, for any real x and y and for $0 \leq t \leq 1$, then $f((1-t)x + ty) \geq (1-t)f(x) + tf(y)$. Then for any set $\{x_1, x_2, \ldots, x_n\}$ of real numbers and any set of positive reals $\{\lambda_1, \lambda_2, \ldots, \lambda_n\}$ with $\lambda_1 + \lambda_2 + \cdots + \lambda_n = 1$, we have that

$$f(\lambda_1 x_1 + \lambda_2 x_2 + \cdots + \lambda_n x_n) \geq \lambda_1 f(x_1) + \lambda_2 f(x_2) + \cdots + \lambda_n f(x_n).$$

A twice differentiable function on an open interval whose second derivative is nonpositive is concave.

4. Sequences and Series

4.1. Convergence of monotone sequences. A bounded monotone (either increasing or decreasing) sequence converges.

4.2. Liminf and Limsup. Suppose that $\{x_n\}$ is a real infinite sequence. Then $a = \limsup_{n \to \infty} x_n$ if and only if, given $\epsilon > 0$, there are infinitely many values of n for which $x_n > a - \epsilon$ and there is an index N for which $x_n < a + \epsilon$ for $n \geq N$.

Also, $a = \liminf_{n \to \infty} x_n$ if and only if, given $\epsilon > 0$, there are infinitely many values of n for which $x_n < a + \epsilon$ and there is an index N for which $x_n > a - \epsilon$ for $n \geq N$.

4.3. Abel's Partial Summation Formula. :

$$\sum_{k=1}^{n} x_k y_k = (x_1 + x_2 + \cdots + x_n)y_{n+1} + \sum_{k=1}^{n}(x_1 + x_2 + \cdots + x_k)(y_k - y_{k+1}).$$

5. Analysis

5.1. Basic theorems. Let $f(x)$ be a continuous real-valued function on a real closed integer $[a, b]$. The *Intermediate Value Theorem* asserts that, within the interval $[a, b]$, the function assumes every value between $f(a)$ and $f(b)$. The *Extreme Value Theorem* asserts that $f(x)$ is bounded on $[a, b]$ and assumes both its least upper bound and greatest lower bound in the interval.

Suppose, further, that $f(x)$ is differentiable on the open interval (a, b). The *Mean Value Theorem* states that there is a number $c \in (a, b)$ for which

$$f'(c) = \frac{f(b) - f(a)}{b - a}.$$

A special case is *Rolle's Theorem*, which provides that if $f(a) = f(b)$, then there is a number $c \in (a, b)$ for which $f'(c) = 0$. If $f(x)$ is n times differentiable on an open interval containing a, then *Taylor's Theorem* asserts that, if $a + u$ belongs to the interval, there is a number v between a and $a + u$ for which

$$f(a + u) = \sum_{k=0}^{n+1} \frac{f^{(k)}(a)}{n!} u^k + \frac{f^{(n+1)}(v)}{(n+1)!} u^{k+1}.$$

The last term, called the *Remainder*, can be written in a number of alternative ways.

If two real-valued functions $f(x)$ and $g(x)$ are defined and differentiable on a neighbourhood of a, if $\lim_{x \to a} f(x) = \lim_{x \to a} g(x) = 0$ and $\lim_{x \to a} (f'(x)/g'(x)) = c$, then, according to *l'Hôpital's Rule*, $\lim_{x \to a} (f(x)/g(x)) = 0$.

If $f(x)$ is a continuous real-valued functions defined on a closed interval $[a, b]$, then the *Integral Mean Value Theorem* asserts that there is a point $c \in [a, b]$ for which $f(c)(b - a) = \int_a^n f(x)\, dx$.

5.2. Fixpoint. Let $f : S \to S$ be a function taking a set into itself. A *fixpoint* of f is an element $s \in S$ for which $f(s) = s$.

5.3. Continuity of monotone functions. Let f be a monotonic (always increasing or always decreasing) real-valued function defined on an open interval I. Then for each $a \in I$, both the one-sided limits $\lim_{x \downarrow a} f(x)$ and $\lim_{x \uparrow a} f(x)$ exist. If they are equal, $f(x)$ is continuous at a. If they are unequal, the $f(x)$ has a *jump discontinuity* at a. Every discontinuity of a monotonic function is of this type.

5.4. Limits involving logarithms.

$$\lim_{t \downarrow 0} t \log t = 0$$

$$\lim_{t \downarrow 0} \frac{\log(1 + t)}{t} = 1$$

6. Linear Algebra

6.1. Linear independence; basis. A set $\{x_1, x_2, \ldots, x_n\}$ of vectors is *linearly independent* if and only if the condition $c_1 v_1 + c_2 v_2 + \cdots + c_n v_n = 0$ for scalars $c_1, c_2, \cdots, c_n$ implies that $c_1 = c_2 = \cdots = c_n = 0$. The *span* of a set $\{v_1, v_2, \cdots, v_n\}$ of a set of vectors, denoted by $\langle v_1, v_2, \cdots, v_n \rangle$ is the set of all linear combinations $c_1 v_1 + c_2 v_2 + \cdots + c_n v_n = 0$ for scalars c_i. A *basis* of a vector space is a linearly independent set whose span is the entire space. The *dimension* of a vector space is the number of elements in any basis.

6.2. Matrix and vector conventions. The action of a matrix on a vector is indicated by the vector in a column appearing to the right of a matrix. Multiplication of matrices proceeds from left to right.

6.3. Linear transformations. A linear transformation T from a vector space V to a vector space W is a mapping that satisfies the condition $T(ax + by) = aT(x) + bT(y)$ for any vectors x and y in V and any scalars a and b. The *kernel* of a linear transformation on a vector space is the set of vectors carried to 0 by the transformation. The dimension of the kernel is the *nullity* of the transformation. The *rank* of the transformation is the dimension of the space of vectors onto which it maps.

6.4. Transpose. The *transpose* of a $m \times n$ matrix $M = (a_{ij})$ is the $n \times m$ matrix whose (i,j)th element is a_{ji}, where $1 \le i \le m$, $1 \le j \le n$. The transpose of M is denoted by M^{t}. A *symmetric matrix* is one that is equal to its transpose.

6.5. Determinant and adjugate matrix. Let $M = (m_{i,j})$ be a $n \times n$ square matrix. The *determinant* of M, denoted by $\det M$ or $|M|$, is equal to the sum of $n!$ terms

$$\sum (-1)^{\epsilon(\sigma)} m_{1,\sigma(1)} m_{2,\sigma 2} \cdots a_{i,\sigma(i)} \cdots a_{n,\sigma(n)}$$

where the sum is over all permutations σ of the set $\{1, 2, \ldots, n\}$ and $\epsilon(\sigma)$ equals 1 when σ can be written as the product of an even number of transpositions and -1 when σ can be written as the product of an odd number of transpositions. For each i with $1 \le i \le n$, the determinant of M can be written in the form

$$\det M = \sum_{j=1}^{n} m_{ij} M_{ij} = \sum_{j=1}^{n} m_{ji} M_{ji},$$

where the number M_{ij} is the *cofactor* of M_{ij}. The *adjugate matrix*, denoted adj$M$, for M has as its (i,j) entry M_{ji}; i.e., it is the transpose of the matrix of cofactors of M. We note that

$$M \cdot \mathrm{adj} M = \mathrm{adj} M \cdot M = (\det M) I.$$

6.6. Invertible and noninvertible matrices. A square matrix M is *nonsingular* or *invertible* if there exists an inverse matrix, denoted by M^{-1} for which $MM^{-1} = M^{-1}M = I$, where I is the *identity* matrix with ones down the main diagonal and zeros elsewhere. Otherwise, it is *singular* or *noninvertible*. The matrix M is nonsingular if and only if its determinant $\det M = |M|$ does not vanish. The nullity of a nonsingular $n \times n$ matrix is 0, and its rank is n.

6.7. Orthogonal matrix. A real square matrix M is *orthogonal* if and only if the product $MM^t = I$ with its transpose is the identity matrix.

6.8. Eigenvectors and eigenvalues. Let M be a $n \times n$ square matrix. An *eigenvector* of M is a nonzero column vector x for which $Mx = \lambda x$ for some scalar λ. Such a scalar is called an *eigenvalue*. The eigenvalues are the roots of the *characteristic polynomial* $\det(\lambda I - M)$ of the matrix. The (algebraic) *multiplicity* of an eigenvalue is equal to its multiplicity as a root of the characteristic polynomial. The *trace* of M is the sum of its eigenvalues (counting multiplicity) and is equal to the sum of its main diagonal elements. The determinant of M is equal to the product of its eigenvalues.

6.9. Minimal polynomial of a matrix. The *minimal polynomial* $p(t)$ of a matrix M is the monic nontrivial polynomial of lowest degree that satisfies $p(M) = O$. The minimal polynomial of a matrix divides its characteristic polynomial.

The *companion matrix* of a monic polynomial $p(t) = t^n + a_{n-1}t^{n-1} + \cdots + a_1 t + a_0$ is the $n \times n$ matrix

$$\begin{pmatrix} 0 & 1 & 0 & 0 & \cdots & 0 \\ 0 & 0 & 1 & 0 & \cdots & 0 \\ 0 & 0 & 0 & 1 & \cdots & 0 \\ & & & \cdots & & \\ -a_0 & -a_1 & -a_2 & -a_3 & \cdots & -a_{n-1} \end{pmatrix}$$

whose characteristic and minimal polynomials are both equal to $p(t)$.

6.10. Jordan canonical form. For any $n \times n$ matrix n, there exists a nonsingular matrix S for which $S^{-1}MS = J$, where the upper triangular matrix J is its Jordan canonical form, where the eigenvalues, repeated according to their multiplicities, occur down the main diagonal. The Jordan canonical form consists of square blocks B_i down its diagonal

$$\begin{pmatrix} B_1 & O & O & O & \cdots & O \\ O & B_2 & O & O & \cdots & O \\ O & O & B_3 & O & \cdots & O \\ & & & \cdots & & \\ O & O & O & O & \cdots & B_k \end{pmatrix}$$

where each block has the form

$$\begin{pmatrix} \lambda & 1 & 0 & 0 & 0 & \cdots & 0 & 0 \\ 0 & \lambda & 1 & 0 & 0 & \cdots & 0 & 0 \\ 0 & 0 & \lambda & 1 & 0 & \cdots & 0 & 0 \\ & & & \cdots & & & & \\ 0 & 0 & 0 & 0 & 0 & \cdots & \lambda & 1 \\ 0 & 0 & 0 & 0 & 0 & \cdots & 0 & \lambda \end{pmatrix}.$$

6.11. Hermitian transpose. For any $m \times n$ matrix M with complex entries m_{ij}, the *hermitian transpose* is the $n \times m$ matrix M^* obtained by taking the complex conjugate of the entries of M and transposing; thus, $M^* = \overline{M^t}$ and the $(i,j)$the element of M^* is $\overline{m_{ji}}$. A matrix A for which $A = A^*$ is said to be *hermitian*.

For the complex column vector x with ith entry x_i, x^* is a row vector whose ith entry is $\bar{x}_i$. The inner product, $\langle x, y \rangle$, of two column vectors is $\sum \bar{x}_i x_j = x^* y$. For matrix A and column vectors x, y, we have that $\langle x, Ay \rangle = \langle A^* x, y \rangle$.

6.12. Three-dimensional vectors. For any vectors $\mathbf{a}$, $\mathbf{b}$ of the real 3-dimensional linear space $\mathbb{R}^3$, we have the *scalar product*

$$\mathbf{a} \cdot \mathbf{b} = a_1 b_1 + a_2 b_2 + c_1 c_2 = |\mathbf{a}| \cdot |\mathbf{b}| \cdot \cos \theta$$

and the *vector product*

$$\mathbf{a} \times \mathbf{b} = (a_2 b_3 - a_3 b_2)\mathbf{i} + (a_3 b_1 - a_1 b_3)\mathbf{j} + (a_1 b_2 - a_2 b_1)\mathbf{k}$$
$$= |\mathbf{a}| \cdot |\mathbf{b}| \cdot (\sin \theta)\mathbf{c}$$

where θ is the angle between the vectors $\mathbf{a}$, and $\mathbf{b}$, $\mathbf{i} = (1,0,0)$, $\mathbf{j} = (0,1,0)$, $\mathbf{k} = (0,0,1)$, and $\mathbf{c}$ is a vector orthogonal to both $\mathbf{a}$ and $\mathbf{b}$. We have the following properties:

(1) $\mathbf{i} \times \mathbf{j} = \mathbf{k}$; $\mathbf{j} \times \mathbf{k} = \mathbf{i}$; $\mathbf{k} \times \mathbf{i} = \mathbf{j}$;
(2) $\mathbf{a} \times \mathbf{b} = -\mathbf{b} \times \mathbf{a}$;
(3) $\mathbf{a} \times (\mathbf{b} \times \mathbf{c}) = (\mathbf{a} \cdot \mathbf{c})\mathbf{b} - (\mathbf{a} \cdot \mathbf{b})\mathbf{c}$;
(4) $\mathbf{a} \cdot (\mathbf{b} \times \mathbf{c}) = \mathbf{b} \cdot (\mathbf{c} \times \mathbf{a}) = \mathbf{c} \cdot (\mathbf{a} \times \mathbf{b})$

$$= \det \begin{pmatrix} a_1 & b_1 & c_1 \\ a_2 & b_2 & c_2 \\ a_3 & b_3 & c_3 \end{pmatrix}.$$

7. Geometry

7.1. Ptolemy's Theorem. The sum of the products of the lengths of pairs of opposite sides of a concyclic quadrilateral is equal to the product of the lengths of the diagonals.

7.2. Sylvester's Theorem. Suppose that $n \geq 2$ points are given in the plane, not all collinear. Then there exists a line that contains exactly two of them. To see this, suppose that n points are given. Pick a point P and line m through two points of the set that does not contain P and for which the distance from P to the line is minimum. If m contains exactly two points of the set, then the result is established. Otherwise, let T be the foot of the perpendicular from P to m and select two points Q and R on m on the same side of T with Q between T and R (Q could coincide with T). Let S be the foot of the perpendicular from Q to the line PR. Since the right triangles

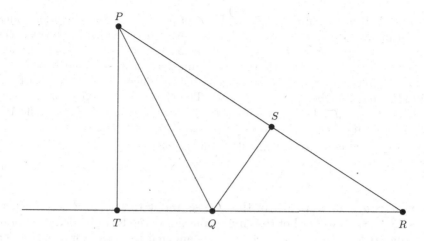

Fig. A.1. Sylvester's Theorem

PTR and QSR are similar and the hypotenuse PS of the first triangle is longer than TR, and so larger than QR, PT must be longer than QS. Thus the distance between Q and PR is smaller than the distance between P and m, contradicting our choice of P and m. (See Fig. A.1.) The result follows.

7.3. Convex set. A *convex set* is any set in the plane or in space that contains along with any two of its points, all points within the line segment that connects them.

7.4. Heron's formula for the area of a triangle. Let the sides of a triangle be a, b, c and let its semiperimeter $\frac{1}{2}(a + b + c)$ be s. Then the area F of the triangle is given by

$$F = \sqrt{s(s - a)(s - b)(s - c)}.$$

We have that

$$16F^2 = 2(a^2b^2 + b^2c^2 + c^2a^2) - (a^4 + b^4 + c^4).$$

7.5. Parabolas. Let a point F (the *focus*) and a line d (the *directrix*) not containing F be given in the plane. The locus of points in the plane that are equidistant from the point F and the line d is called a *parabola*. All parabolas in the plane are similar. The graph of any equation of the form $y = ax^2 + bx + c$ with $a \neq 0$ is a parabola. One characteristic of a parabola is its *reflection property*. For any point P on the parabola, the reflection of the line FP about the normal to the parabola at P is a line parallel to its axis.

7.6. Ellipses. An *ellipse* is the locus of points P in the plane for which the distance PF from P to a fixed point F (the *focus*) divided by the distance from P to a fixed line d (the *directrix*) is a constant $e \in (0, 1)$. This constant is called the *eccentricity*. Alternatively, an ellipse can be described as the locus of points P for which the sum of the distances from P to two fixed points F_1 and F_2 (the *foci*) is constant. The ellipse has a *reflection property* in that, for any point P on the ellipse, the line PF_1 and PF_2 are reflected images of each other with respect to the normal to the ellipse at P.

A standard equation for an ellipse in the plane is

$$\frac{x^2}{a^2} + \frac{y^2}{b^2} = 1,$$

where $a > b > 0$. The foci of this ellipse are at the points $(\pm c, 0)$, where $c = \sqrt{a^2 - b^2}$. The chord of the ellipse containing the foci is called the *major axis*, and its length is $2a$. The chord perpendicular to the major axis is called the *minor axis*, with length $2b$. The eccentricity of this ellipse is $e = c/a$ and the directrices for the two foci are the vertical lines with equations $x = \pm a^2/c$.

7.7. Hyperbolas. A *hyperbola* is the locus of points P in the plane for which the distance PF from P to a fixed point F (the *focus*) divided by the distance from P to a fixed line d (the *directrix*) is a constant e (the eccentricity) exceeding 1. Alternatively, a hyperbola can be described as the locus of points P for which the difference of the distances from P to two fixed points F_1 and F_2 is constant. A standard equation for a hyperbola in the plane is

$$\frac{x^2}{a^2} - \frac{y^2}{b^2} = 1,$$

where $a > b > 0$. This curve is symmetric about each of the coordinate axes, is unbounded and, for large values of $|x|$, approaches its *asymptotes*, the two lines $y = \pm (b/a)x$.

8. Group Theory

8.1. Definition of group. A *group* G is a set with an associative binary operation $(x, y) \longrightarrow xy$ that contains an identity element e for which $e^2 = e$ and $ue = eu = u$ for every element $u \in G$. For each element $a \in G$, there exists an inverse element v for which $uv = vu = e$.

8.2. Subgroups and cosets. A *subgroup* H of a group G is a group with the same binary operation as in G. A *left coset* of H in G is a set of the form $xH = \{xh : h \in H\}$; a *right coset* has the form $Hx = \{hx : h \in H\}$. A subgroup H of G is *normal* iff $xH = Hx$ for all $x \in G$, or, equivalently, $x^{-1}hx \in H$ whenever $h \in H$.

8.3. Order. The *order* of a subgroup H of G is the number of elements in H and the order of an element $x \in G$ is the smallest positive integer m for which $x^m = e$. *Lagrange's Theorem* provides that the order of each subgroup and of each element divides the order of the group. The *index* of a subgroup of a group is equal to the order of the group divided by the order of the subgroup.

8.4. Homomorphism. A *homomorphism* of a group G to a group K is a mapping $\phi : G \to K$ for which $\phi(xy) = \phi(x)\phi(y)$ and $\phi(x^{-1}) = (\phi(x))^{-1}$ for all $x, y \in G$. The *kernel* of the homomorphism ϕ is the set of elements of G that get mapped to the identity of K. The kernel of a homomorphism is a normal subgroup of G. Conversely, every normal subgroup H of G is the kernel of a homomorphism of G onto some other group, in particular the factor group G/H whose elements are the cosets of H.

8.5. Automorphism. An *automorphism* of a group G is a one-one mapping ϕ of G onto itself for which $\phi(e) = e$, $\phi(x^{-1}) = (\phi(x))^{-1}$ and $\phi(xy) = \phi(x)\phi(y)$ for all $x, y \in G$.

8.6. Permutations. A *permutation* is a one-one mapping of the set $\{1, 2, \ldots, n\}$ onto itself. The *symmetric group* S_n is the group of all $n!$ permutations on the first n positive integers with the operation of functional composition. A transposition is a permutation that switches precisely two numbers and leaves the rest fixed; the transposition that takes a to b and b to a is denoted by (ab). If x is any permutation that takes a to c and b to d, then $x^{-1}(ab)x = (cd)$. A permutation is *even* iff it can be written as the product of an even number of transpositions; it is odd, otherwise. The set of all even permutations constitutes the *alternating group* A_n. A cycle is a permutation of the form $(abc\ldots z)$ that takes a to b, b to c, and so on until finally z to a. The order of a cycle is equal to the number of elements affected by the cycle. Each permutation can be written as a product of cycles, with multiplication proceeding from left to right.

9. Combinatorics

9.1. Pigeonhole principle. The *Pigeonhole principle* states that if a finite set of objects is partitioned into a number of non-overlapping categories, and there are more objects than categories, then some category must be assigned at least two objects.

9.2. Graphs. A *graph* consists of a set of points, called *vertices* or *nodes*, along with a set of edges that connect certain pairs of vertices. A vertex may be connected to itself, and two vertices may be connected by more than one edge. If P and Q are two vertices, then the edge joining them is denoted by $P--Q$. A vertex is incident with an edge if it is one of its endpoints. A *path* is a finite sequence of vertices, each consecutive pair of which is joined by an edge. A *closed path* is a path that includes an edge joining the final and initial vertices. A path is a *chain* if it does not contain any vertex more than once. A closed path is a *circuit* or *loop* if it does not encounter any vertex more than once.

9.3. Trees. A *tree* is a graph that contains no loop and for which there is exactly one path between any two vertices. The number of vertices in a tree exceeds the number of edges by 1. A *root* of a tree is any vertex on the tree that is incident with exactly one edge. A *spanning tree* or *path* is one that contains every vertex of the graph.

9.4. Planar graphs. A *planar graph* is a graph that can be represented by vertices and edges drawn on a plane or spherical surface without any edges intersecting. *Faces* are regions surround by loops that contain no vertices. If V, E, F are respectively the number of vertices, edges and faces in a planar graph or a three-dimensional polyhedron, then Euler's equation $V - E + F = 2$ is valid.

9.5. Directed graphs. A *directed graph* is one for which each edge is assigned a direction so that it is conceived as emanating from its first endpoint to its second: $P \to Q$. A (directed) path or tree in a directed graph is such that the exit vertex of each edge is the entry vertex of the next: $P \to Q \to R$. A root of a spanning path or tree is a vertex from which each other vertex can be reach along a directed path.

10. Number Theory

10.1. Base of numeration. Let b be a positive integer exceeding 1. Every positive integer a can be written in the form

$$a = a_r b^n + a_{n-1} b^{n-1} + \cdots + a_1 b + a_0$$

where r is a positive integer and the coefficients a_k are integers for which $0 \le a_k \le b - 1$. Such a representation, denoted by

$$\overline{a_r a_{r-1} \ldots a_1 a_0},$$

is the base b numeration of a.

10.2. Greatest common divisor and least common multiple. The *greatest common divisor* of two integers a, b, denoted by gcd (a, b), is the largest positive integer that divides both of them. Given any pair (a, b) of integers, there exist integers x and y for which gcd $(a, b) = ax + by$. Two integers are coprime if their greatest common divisor is 1. The *least common multiple* of a and b, denoted by lcm (a, b), is the smallest positive integer that both of them divide into. For any pair (a, b) of integers, $ab = \gcd(a, b) \cdot \text{lcm}(a, b)$.

Let a and b be integers whose greatest common divisor is g. Then there exist integers x and y for which $ax + by = g$. As x and y vary, $ax + by$ runs through all the multiples of g.

10.3. Modularity and residues. Let m be a positive integer exceeding 1, and let a and b be arbitrary integers. We say that a is congruent to b modulo m iff $a - b$ is a multiple of m; in symbols, $a \equiv b \pmod{m}$. If a and m are coprime, and b is arbitrary, then there always exists an integer x that satisfies the congruence $ax \equiv b \pmod{m}$. A *complete set of residues* modulo m is any finite set of integers that contains an integer congruent to each i with $0 \le i \le m - 1$. In particular, when a and m are coprime, $\{ak : 0 \le k \le m - 1\}$ is a complete set of residues.

10.4. Finite number fields. Let p be a prime number. The set $\mathbb{Z}_p$ is the field whose elements are $\{0, 1, 2, \ldots, p - 1\}$, where addition and multiplication are defined modulo p. The set $\mathbb{Z}^*$ consisting of $\{1, 2, \cdots, p - 1\}$ constitutes a group under multiplication. With these operations of addition and multiplication, $\mathbb{Z}_p$ is a field.

10.5. The divisor function. Let n be a positive integer and $\tau(n)$ be the number of its positive divisors. If m and n are coprime, then $\tau(mn) = \tau(m)\tau(n)$. If $n = \prod p^a$ where the product is taken over all prime divisors of n, then $\tau(n) = \prod(a + 1)$.

10.6. Fermat's Little Theorem. Let p be a prime and a an integer not divisible by p. Then $a^{p-1} \equiv 1 \pmod{p}$.

APPENDIX B

Top Ranking Students

Generally, prizes were awarded to the top three to five students in the contest. Because the top grades were closer to each other in some years than in others, the way in which the students were recognized was not consistent from year to year.

Each problem was marked out of 10. After the first few contests, to discourage students from frittering their efforts over too many problems, only the five highest marks for the problems were counted unless the total of these marks exceeded 30, in which case the scores for all the problems were included in the total.

Competition 1: March 18, 2001 (13 candidates)
1. Jonathan Sparling
2. Jimmy Chui
3. Pavel Gyrya

Honourable mention: Ari Brodsky, Nicholas Martin, Al Momin

Competition 2: March 9, 2002 (11 candidates)
1. Jimmy Chui
2. David Varodayan
3. Isaac Li
4. Fred Dupuis; Robert Ziman (tied)
6. Emily Redelmeier

Competition 3: March 16, 2003 (20 candidates)
1. Jonathan Sparling
2. David Varodayan
3. Garry Goldstein
4. Benjamin Moull
5. Jimmy Chui

Honourable Mention: Samuel Huang, Emily Redelmeier, Robert Ziman

© Springer International Publishing Switzerland 2016
E.J. Barbeau, *University of Toronto Mathematics Competition (2001–2015)*, Problem Books in Mathematics,
DOI 10.1007/978-3-319-28106-3

Competition 4: March 14, 2004 (18 candidates)

1. Garry Goldstein
2. Ali Feiz Mohammadi
3. Robert Barrington Leigh
4. Roger Mong

Honourable mention: Samuel Huang, Emily Redelmeier, David Shirokoff, Ilya Sutskever

Competition 5: March 12, 2005 (21 candidates)

1. Robert Barrington Leigh
2. Jacob Tsimerman
3. Garry Goldstein

Honourable mention: Ali Feizmohammadi, Tianyi David Han, Samuel Huang, Emily Redelmeier, Ilya Sutskever

Competition 6: March 12, 2006 (17 candidates)

1. Jacob Tsimerman
2. Robert Barrington Leigh
3. Janos Kramar
4. Garry Goldstein
5. Ali Feizmohammadi

Competition 7: March 11, 2007 (10 candidates)

1. Janos Kramar
2. Mu Cai
3. Geoffrey Siu

Honourable mention: Zhiqiang Li, Samuel Huang, Elena Flat

Competition 8: March 9, 2008 (14 candidates)

1. Janos Kramar
2. Konstantin Matveev
3. Sida Wang
4. Oleg Ivrii

Honourable mention: Viktoriya Krakovna

Competition 9: March 8, 2009 (21 candidates)

1. Konstantin Matveev
2. Alexander Remorov
3. Viktoriya Krakovna
4. Junjiajia Long
5. Sergio DaSilva

Competition 10: March 7, 2010 (31 candidates)

1. Alexander Remorov
2. Sida Wang
3. Keith Ng

4. Viktoriya Krakovna
5. Sergei Sagatov

Competition 11: March 6, 2011 (26 candidates)

1. Keith Ng; Alexander Remorov (tied)
3. Sergei Sagatov
4. Philip Chen
5. Fengwei Sun
6. Jungwei Huo
7. Mengdi Hua

Competition 12: March 10, 2012 (38 candidates)

1. Jonathan Zung
2. Yu Wu
3. Keith Ng
4. Jialin Song
5. Jonathan Love

Competition 13: March 9, 2013 (38 candidates)

The top three students are listed in alphabetical order as their scores were very close:

Jonathan Love
Jialin Song
Jonathan Zung

The next three ranking students also listed in alphabetical order were:

RongXi (Bill) Guo
Shia Xiao Li
Ruize Luo

Competition 14: March 9, 2014 (31 candidates)

1. Jonathan Love
2. Brian Bi
3. Shijie Xiu
4. Yizhou Sheng
5. David Perchersky

Competition 15: March 8, 2015 (33 candidates)

1. Jonathan Love
2. Shijie Xiu
3. Michael Chow
4. Chengyuan Yao
5. Jerry Jia Wang
6. Gal Gross

Since 2007, the contest papers have been set in memory of Robert M. Barrington Leigh (1986–2006), a brilliant and promising mathematics student from Edmonton, Alberta, whose untimely death in August, 2006 occasioned great sorrow among his friends, mentors and professors.

Index

© Springer International Publishing Switzerland 2016
E.J. Barbeau, *University of Toronto Mathematics Competition
(2001–2015)*, Problem Books in Mathematics,
DOI 10.1007/978-3-319-28106-3